11th European Summer School on Experimental Nuclear Astrophysics 2022

EPJ Web of Conferences Volume 275 (2023)

Catania, Italy
12 – 19 June 2022

Editors:

L. Lamia
R.G. Pizzone
G.G. Rapisarda
M.L. Sergi

ISBN: 978-1-7138-7049-4

Printed from e-media with permission by:

Curran Associates, Inc.
57 Morehouse Lane
Red Hook, NY 12571

Some format issues inherent in the e-media version may also appear in this print version.

No changes have been made to the content of these proceedings. There may be changes to pagination and minor adjustments for aesthetics.

Printed with permission by Curran Associates, Inc. (2023)

For additional information, please contact EDP Sciences – Web of Conferences at the address below.

EDP Sciences – Web of Conferences
17, Avenue du Hoggar
Parc d'Activité de Courtabœuf
BP 112
F-91944 Les Ulis Cedex A
France

Phone: +33 (0) 1 69 18 75 75
Fax: +33 (0) 1 69 28 84 91

contact-edps@webofconferences.org

Additional copies of this publication are available from:

Curran Associates, Inc.
57 Morehouse Lane
Red Hook, NY 12571 USA
Phone: 845-758-0400
Fax: 845-758-2633
Email: curran@proceedings.com
Web: www.proceedings.com

TABLE OF CONTENTS

THE 11TH EUROPEAN SUMMER SCHOOL ON EXPERIMENTAL NUCLEAR ASTROPHYSICS

Primordial Nucleosynthesis and Early Stars, Stellar Evolution, Hydrostatic and Explosive Nucleosynthesis, Plasmas in Stars and Laboratories, Detectors and Facilities for Nuclear Astrophysics, Experiments with rare and radioactive isotopes, Indirect Methods

SCIENTIFIC COMMITTEE

M. Aliotta, C. Bertulani, M. Busso, E. Chavez, A. Coc, M. El Eid, T. Kajino, K.L. Kratz, S. Kubono, K. Langanke, J. Josè, T. Motobayashi, O. Straniero, G. Rogachev, S. Romano, C. Spitaleri, L. Trache, R.E. Tribble, W. Liu, M. Wiescher

LOCAL COMMITTEE

R.G. Pizzone (Chair), L. Lamia (co-Chair), G. D'Agata, G.L. Guardo, M. La Cognata, D. Lattuada, A.A. Oliva, R. Spartà, A. Tumino

SCIENTIFIC SECRETARY

G.G. Rapisarda and M.L. Sergi
astro2022@lns.infn.it

SECRETARY V. Potenza, C. Di Mauro **CONTACT MEDIA** G. Agnello

Università di Catania · Uni ct FISICA E ASTRONOMIA "ETTORE MAJORANA" · INFN LNS · CHETEC INFRA edp sciences

EPJ Web of Conferences

Statement of Peer review

In submitting conference proceedings to EPJ Web of Conferences, I certify to the Publisher that I adhere to the **Policy on Publishing Integrity** of the journal in order to safeguard good scientific practice in publishing.

1. All articles have been subjected to peer review administered by the proceedings editors.
2. Reviews have been conducted by expert referees, who have been requested to provide unbiased and constructive comments aimed, whenever possible, at improving the work.
3. Proceedings editors have taken all reasonable steps to ensure the quality of the materials they publish and their decision to accept or reject a paper for publication has been based only on the merits of the work and the relevance to the journal.

Title, date and place of the conference Catania, 12-19 June 2022

11th European Summer school on Experimental Nuclear Astrophysics

Proceedings editor(s): L. LAMIA, R.G. PIZZONE, G.G. RAPISARDA, M.L. SERGI

Date and editor's signature 30.11.2022

17, avenue du Hoggar - PA de Courtabœuf – BP 112 - 91944 Les Ulis Cedex A (France)
Tél. : 33 (0)1 69 18 75 75 - Fax : 33(0)1 69 07 45 17 – www.edpsciences.org

Preface

In this book a collection of the lecture notes given during the Eleventh European Summer School on Experimental Nuclear Astrophysics is reported. The school belongs to the European Network of Nuclear Astrophysics Schools (ENNAS), made by the European schools on nuclear astrophysics and related areas (the Santa Tecla, the Sinaia, the Russbach schools) and having the common effort of preparing and educating young physicists in nuclear physics, astrophysics and their mutual relationship in the nuclear astrophysics field. The school, whose first edition was held in 2001, took place from 12 to 19 of June 2022 in Catania. The School was hosted by the INFN Laboratori Nazionali del Sud, one of the 4 INFN national laboratories whose main activity is focused in Nuclear and particle astrophysics. A total number of 70 participants between young students and researchers from more than 20 countries of all the world attended the lectures. Due to the spirit of the school, young researchers and students were strongly encouraged to present their latest works and results.. Lectures were given by twenty of the best top scientists in the field.

During the Covid19 pandemic many of the traditional features of academic learning and education were forcedly changed. Ph. D. students, were not given the possibility to travel for school or conferences, despite the large efforts by the scientific community to have online events. Therefore, at the eve of this Eleventh school, we are particularly proud. We continue our commitment for students education despite the pandemic and without any discrimination, allowing students of less favoured countries to attend a top-scale academic environment.

The school has tried once more to present to the young students the global picture of nuclear astrophysics research in the last years. Thus the scientific program of the school covered a wide range of topics dealing with various aspects of nuclear astrophysics, such as stellar evolution and nucleosynthesis, Big Bang nucleosynthesis, direct and indirect methods, radioactive ion beams. Nuclear astrophysics plays a key role for understanding energy production in stars, stellar evolution and the concurrent synthesis of the chemical elements and their isotopes. It is also a fundamental tool to explain the ashes of the early universe, to determine the age of the universe through the study of pristine stellar objects and to predict the evolution for the Sun or Stars. The bone structure for the above aspects is based on nuclear reactions, whose rates need to be determined in laboratories. Although impressive progress has been made over the past decades, which was rewarded by Nobel prizes, several open questions are still unsolved, which challenge the basis of the present understanding.

A list of the lecture topics is given below:

- Big Bang Nucleosynthesis;
- Stellar nucleosynthesis;
- Plasmas in stars and laboratories;
- Radioactive ion beams;
- Neutron star mergers and r-process;

- Detector and facilities for nuclear astrophysics;
- Direct and Indirect methods in nuclear astrophysics;

An often quoted statement of Carl Sagan, "We are all star stuff", is exhaustively describing one of the great discoveries of the XX century as well as the main aim of nuclear astrophysics. The present theory of stellar nucleosynthesis (based on the famous paper B2FH) predicts chemical evolution of the Universe, which is testable by looking at stellar spectral lines, meteorites, pre- solar grains or other means of investigation. Big bang nucleosynthesis tells us that the early universe consisted of only the light elements, and so one expects the first stars to be composed of hydrogen, helium, and lithium, the three lightest elements. The achievements of WMAP and Planck missions, with the precise measurement of many cosmological parameters,have re-ignited the interest on the primordial nucleosynthesis, especially for the still unclear Lithium primordial abundance. Stellar structure and the H-R diagram indicate that the lifetime of a star depends greatly on its initial mass and chemical composition, so that massive stars are very short-lived, and less massive stars are longer-lived. As a star dies, nuclear astrophysics argues that it will enrich the interstellar medium with heavy elements (in this case all elements heavier than lithium, the third element), from which new stars are formed. This account is consistent with the observed correlation between stellar metallicity and red shift.

The theory of stellar nucleosynthesis would not be very convincing if reliable nuclear physics inputs are adopted. By carefully scrutinizing the table of nuclides, nuclear astrophysicists were able to predict the existence of different stellar environments which could produce the observed isotopic abundances, and the nuclear processes which must occur in these stars. This is also valid for elements heavier than iron production, when fusion reactions are of negligible importance and the main nucleosynthetic path goes through p-process, r-process, and s-process. Many of the nuclides generated in the huge explosion triggered by the last burnings in the life of a massive star are unstable. Then in order to understand what is going on in Supernovae explosions, Gamma-ray bursts, Novae it was pointed out that the physics of radioactive beams should be explored. Many group of physicists have then taken the opportunities offered by the world-wide developments of radioactive ion beams facilities to start to investigate such environments. In this field also the weak interaction processes are important and in some cases, like type II Supernovae, dominant. This is extremely important in the multi-messenger astronomy era.

In order to better understand our Universe, a precise knowledge of nuclear inputs for nuclear astrophysics is therefore needed. They influence sensitively the nucleosynthesis of the elements in the early stages of the universe and in all the objects formed thereafter and control the associate energy generation, neutrino luminosity and stellar evolution. Thus a good knowledge of reaction rates is essential to understand this broad picture. Studies at the energies typical of nuclear astrophysics for charged particle induced reactions (few keV - 1 MeV) are severely hampered by the low signal-to-noise ratio, essentially due to the tiny cross sections that have to be measured (as low as picobarn in some cases). The best experimental solutions to this main problem will be discussed in this book and presented in their details. Selected experimental techniques such as underground facilities, recoil separators, spectrometers will be reviewed together with the wide field of nuclear astrophysics research opened by the development of radioactive ion beams in several laboratories of the world.

Detectors and facilities useful for the next decade of nuclear astrophysics studies were reviewed in the lectures and a big attention was devoted to future research possibilities for studies. This knowledge and facilities will be of groundbreaking importance for solving the future quests for this field. Although all these efforts in some cases (such as the $^{12}C+^{12}C$ interaction, the $^{12}C+^{16}O$, the $^{12}C(\alpha,\gamma)^{16}O$ and many others) the goal is far from being achieved by means of direct measurements. Moreover the presence of the electron screening effect makes very hard and sometimes impossible the measurement of the bare nucleus astrophysical factor, that is the quantity needed for astrophysics. Thus extrapolations are often adopted. An alternative way to by-pass these problems and measure the bare nucleus astrophysical factor or cross section is given by the indirect methods. Among them the Coulomb Dissociations, the ANC, and the Trojan Horse Method will be reviewed.

In recent years, new possibilities of exploring the universe have opened. In this view, stars and galaxies are nowadays investigated not only by means of electromagnetic spectra but also through new "eyes", in the framework of the multi-messenger astronomy. In particular, gamma rays and gravitational waves play an important role for a better understanding of what is going on in the universe.

The school has lived a particularly strong moment with the public lecture of prof. Jocelyn Bell at the "Cittá della Scienza", which was attended by 150 applicants (70 school students and lecturers plus 80 attendees who registered online).

We gratefully acknowledge all the participants, the lecturers and the members of the Scientific committee.The school could not have taken place without their efforts and help. The Organizing Committee also acknowledges the financial support of the Laboratori Nazionali del Sud - Istituto Nazionale di Fisica Nucleare - Italy, Dipartimento di Fisica e Astronomia "Ettore Majorana" of the University of Catania - Italy, Centro Siciliano di Fisica Nucleare e Struttura della Materia and Chetec-Infra network. The school has been honoured by the welcome addresses by prof. F. Priolo, rector of the Catania University, by Dr. Santo Gammino, Director of Laboratori Nazionali del Sud, by prof. M.G. Grimaldi, Director of the Dipartimento di Fisica e Astronomia, and by prof. A. Tricomi, Director of INFN-Sezione Catania & Centro Siciliano di Fisica Nucleare e Struttura della Materia. We wish a fruitful reading of this book.

L. Lamia
R.G. Pizzone

Transfer Reactions in Nuclear Astrophysics

Philip Adsley[1,2,*]

[1]Cyclotron Institute, Texas A&M University, College Station, Texas 77843-3366, USA
[2]Department of Physics and Astronomy, Texas A&M University, College Station, Texas 77843-4242, USA

Abstract. Transfer reactions are important tool in nuclear astrophysics. These reactions allow us to identify states in nuclei and to find the corresponding energies, to determine if these states can contribute to astrophysical nuclear reactions and ultimately to determine the strength of that contribution. In this paper, the basic details of how transfer reactions may be used in nuclear astrophysics are set out along with some common pitfalls to avoid.

1 Introduction

Direct measurements of nuclear cross sections at energies relevant to stellar burning are extremely challenging. Indirect methods provide an alternative way of determining astrophysically important reaction rate. Transfer reactions are one of these indirect methods. This paper briefly sets out the background of transfer reactions and their utility in nuclear astrophysics. More detailed discussions are available in other sources such as Refs. [1, 2].

Transfer reactions are, unsurprisingly, reactions in which something is transferred. This transferred item can be a single nucleon (a proton or a neutron), a cluster (such as an α particle or a deuteron) or can include swapping two particles in charge-exchange reactions such as (p, n) or $(^3\text{He}, t)$. Transfer reactions happen quickly, typically on the order of 10^{-22} seconds, and are dominated by peripheral reactions where the nuclear reaction takes place at the surface of the nucleus. A schematic diagram for a transfer reaction is shown in Fig. 1.

In general, to model transfer reactions we used calculations such as the (first-order) Distorted-Wave Born Approximation (DWBA). More complex reaction models such as coupled-channel calculations are sometimes necessary but beyond the current scope; for a comprehensive treatment, see the book by Thompson and Nunes [3]. These calculations themselves require a number of inputs, including experimental values such as the beam and target species, the beam energy, the spins of the various states, the transferred spin and orbital angular momentum etc. There are also other inputs such as optical-model potentials. These potentials describe the interaction between the nuclei in the entrance channel and the nuclei in the exit channel, the binding of transfer fragment to the core systems in both the initial and final partitions, and possibly other interactions such as the core-core interaction.

2 Background

In order to calculate thermonuclear reaction rates in stars the cross section, $\sigma(E\text{c.m.})$, which represents the probability of interaction between two particles with a centre-of-mass energy

*e-mail: padsley@tamu.edu

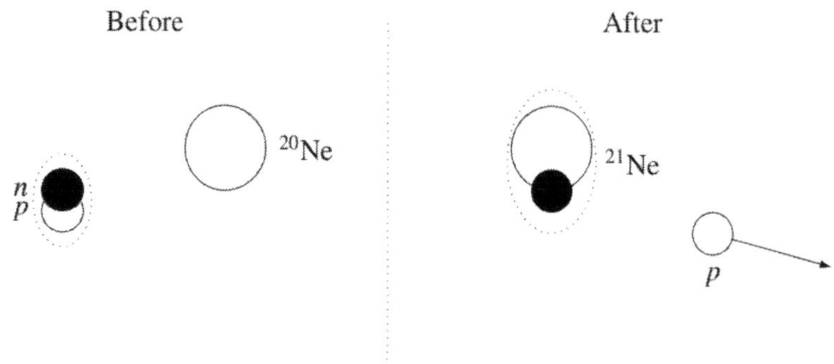

Figure 1. A schematic diagram showing a transfer reaction, in this case ^{20}Ne$(d,p)^{21}$Ne. The deuteron projectile is shown as a neutron (black filled circle) and a proton (small hollow circle) bound together, and the ^{20}Ne target is shown as a large hollow circle.

$E_{\text{c.m.}}$, must be folded with the Maxwell-Boltzmann distribution, representing the likelihood of the two particles having that relative energy. The thermonuclear reaction rate is then found by integrating this product across all energies, representing a sum over all possible energies of the product of the interaction probability with the probability of two particles having that relative energy.

The cross section can have a number of different contributions including direct-capture and resonance reactions. Direct capture is an electromagnetic transition from an initial continuum state in which a projectile is moving outside a nucleus to a final bound state within the nucleus. Since the electromagnetic force is well known, the inputs from nuclear physics are the energy, spin and parity of the bound state, and the overlap between the initial nucleus+projectile configuration and the final bound state in the new residual nucleus. This later quantity is expressed as the spectroscopic factor, C^2S, or the asymptotic normalisation coefficient. The direct-capture cross section can be calculated using $\sigma_{\text{DC}}(E_{\text{c.m.}}) = \sum_i (C^2S)_i \sigma_{i,\text{DC}}(E_{\text{c.m.}})$, where $(C^2S)_i$ is the spectroscopic factor of the ith state and $\sigma_{i,\text{DC}}(E_{\text{c.m.}})$ is the single-particle direct-capture cross section at the centre-of-mass energy E which is calculated analytically using codes such as TEDCA [2, 4].

Resonance reactions are two-step processes - first the incoming projectile fuses with the target nucleus, forming an excited resonance state. Next, this resonance state decays, usually by the emission of another particle or a γ ray. For resonance reactions, the cross section can vary significantly more than for direct capture. The reaction cross section is typically written as in Breit-Wigner form:

$$\sigma_{\text{res}}(E_{\text{c.m.}}) = \omega \frac{\pi \hbar^2}{2\mu E} \frac{\Gamma_{\text{in}}\Gamma_{\text{out}}}{(E_{\text{c.m.}} - E_R)^2 + \Gamma^2/4} \tag{1}$$

where $\omega = \frac{2J+1}{(2j_1+1)(2j_2+1)}$ is the spin-statistical factor with J the spin of the resonance and $j_{1,2}$ the spins of the reactants, μ the reduced mass, $\Gamma_{in(out)}$ the partial width for the entrance (exit) channel, Γ is the total width $\Gamma = \sum_i \Gamma_i$, E_R is the resonance energy and $E_{\text{c.m.}}$ is the centre-of-mass energy at which the reaction takes place.

3 Uses of transfer reactions in nuclear astrophysics

Having covered the background of how nuclei react through different mechanisms, the use of transfer reactions in determining the relevant nuclear quantities of interest is now discussed.

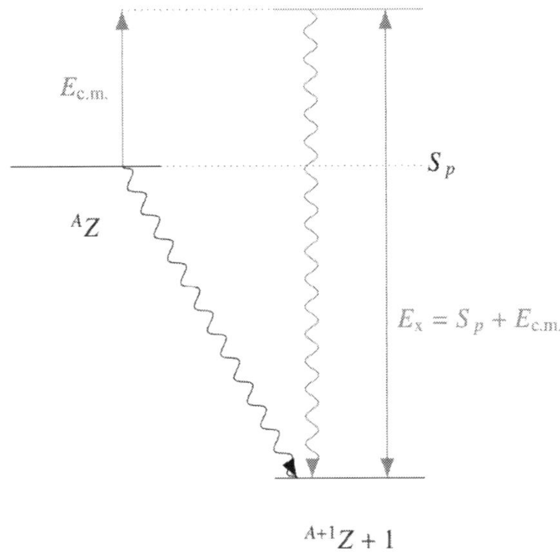

Figure 2. Schematic diagrams showing the (black) direct-capture mechanism and (red) resonance-capture mechanism for proton capture onto a target nucleus ^AZ to the ground state of the residual nucleus, $^{A+1}Z + 1$. The excitation energy of the resonance state of importance for the reaction is given by the sum of the proton separation energy, S_p, and the centre-of-mass energy of the colliding system, $E_{\rm c.m.}$.

From the previous section, it is possible to list the nuclear quantities required to calculate the reaction rate. These include the existence and energies of nuclear states, their spin and parity, and their overlap with various initial and final configurations including various different quantities defined below such as the spectroscopic factor, the asymptotic normalisation coefficient, and the partial width.

One important consideration when using transfer reactions for nuclear astrophysics is that the information gleaned is model-dependent and, if inappropriate models are used to describe the data, then the conclusions drawn are potentially deficient.

3.1 State and resonance energies

The simplest use of transfer reactions is in finding if states exist and, if so, at what energies they occur in the compound nucleus. The energies of states are determined by using two-body kinematics for the transfer reaction and E_x/E_r reconstructed from the beam energy, the known masses and the detected energies and angles of the emitted radiations.

One complication is that transfer reactions tend to be selective; this is sometimes a great advantage but it can be a problem. Many transfer reactions are selective to the states populated and the selectivity may not be to the states which are most important astrophysically. This can lead to bias since only a subset of the states are used to evaluate the reaction rate. Some caution is required in drawing firm conclusions as to the non-existence of states from only a single transfer reaction.

An example of a study in which a transfer reaction has been used to find excited states in a nucleus is ^{50}Cr$(p,t)^{48}$Cr, measured a number of years ago with the K600 magnetic spectrometer and an array of silicon detectors [5–7] at iThemba LABS, South Africa. Many states in ^{48}Cr were observed for the first time, states which have a potential impact on the ^{44}Ti$(\alpha,p)^{47}$V

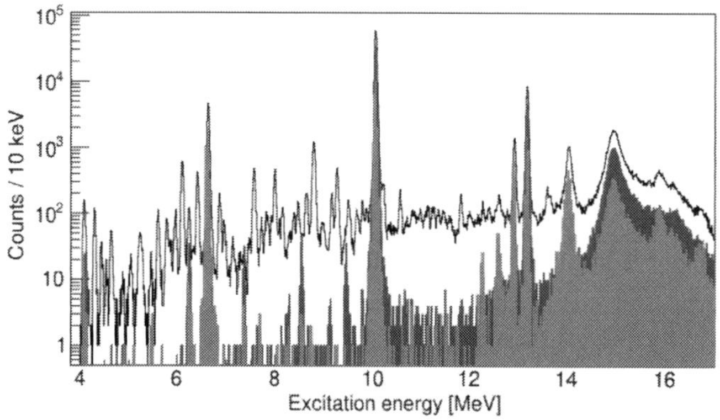

Figure 3. An excitation-energy spectrum of states in ^{48}Cr following the ^{50}Cr$(p,t)^{48}$Cr reaction [7]. The hollow black spectrum corresponds to states in ^{48}Cr while the blue and red spectra show ^{12}C$(p,t)^{10}$C and ^{16}O$(p,t)^{14}$O reactions from carbon and Mylar targets, respectively.

reaction which destroys ^{44}Ti produced in core-collapse supernovae. An excitation-energy spectrum for this reaction is shown in Fig. 3 showing the large number of ^{48}Cr states populated, many observed for the first time, giving valuable information on the accuracy of calculations of the ^{44}Ti$(\alpha,p)^{47}$V reaction rates.

3.2 Spins and parities

Spins and parities can be constrained using transfer reactions. Spins and parities are needed because they determine which possible orbital angular momenta can contribute to astrophysical reactions. An higher orbital angular momentum means a lower penetrability through the barrier for the incoming particle, and a smaller contribution to the reaction rate. From the differential cross section, the probability of having a reaction take place at a particular angle, the transferred angular momentum of a reaction can be determined.

Again, there are complications with this process. One is that many transfer reactions are insensitive to the total spin of the final state, with the gross structures of the differential cross section being determined predominantly from the transferred angular momentum. This means that, for many single-nucleon transfer reactions, the spin of the final state can only be determined to $j = \ell \pm 1/2$. There are some exceptions to this: first, the spin-orbit term of the potential can induce some fine structure which can allow for discrimination between various final spins and, second, transfer reactions in which tensor analysing powers can be measured (e.g., using polarised deuterons in the $\left(\vec{d},p\right)$ reaction) allows for discrimination between different final spins.

3.3 Spectroscopic factors and asymptotic normalisation coefficients

Both direct-capture and resonance-capture reactions rely on the behaviour of the wave function of the final state. For direct capture, what matters is how like the initial configuration the final state looks, described by the spectroscopic factor. For resonance capture, what matters is how likely the transferred particle is to appear at the nuclear surface in the final state; this, in turn, depends on the single-particle wave function and the spectroscopic factor. The likelihood of appearance is then multiplied by the likelihood of the transferred particle making it through the potential barrier for the nucleus (the penetrability) to get the partial width for the state. Due to time-reversal invariance, this partial width is also the likelihood of getting that particle into the nucleus through the potential barrier into the resonance state in question, as needed to calculate the cross section.

The spectroscopic factor, C^2S is an empirical factor which relates the calculated cross section to the observed cross section $\frac{d\sigma}{d\Omega}_{\rm exp} = C^2S\frac{d\sigma}{d\Omega}_{\rm calc}$ where $\frac{d\sigma}{d\Omega}_{\rm exp(calc)}$ is the experimental (calculated) differential cross section.

Spectroscopic factors can be used to calculate the direct-capture reaction component as described in Section 2. For resonance-capture reactions, the quantity of interest is the partial width. The partial width can be written as [1]:

$$\Gamma = 2P_\ell(E,R)\frac{\hbar^2 R}{2\mu}C^2S\,|\phi(R)|^2, \tag{2}$$

where P_ℓ is the penetrability through the barrier with angular momentum ℓ, μ is the reduced mass, and $\phi(R)$ is the radial wave function desribing the relative motion of the transferred nucleon in the final state at the radius R. The spectroscopic factor here (C^2S) is the same as previously. However, the measured spectroscopic factor depends on the potentials chosen to model the reaction. For nuclear astrophysics, it is better to calculate the partial width using the same potentials and radial wave functions as the DWBA calculation and determination of the spectroscopic factor. This will not remove the entire model-dependency of the reaction but the uncertainties in the partial width from the reaction model are significantly reduced by performing self-consistent calculations, see Ref. [10] for an example.

The asymptotic normalisation coefficient (ANC) measures the normalisation of the tail of the nuclear overlap, and are used both in the calculation of direct-capture reactions into bound states and the determination of the partial width for resonances [3]. ANCs are less model-dependent in terms of the connection between the measured transfer cross section and the partial width of interest. However, since ANCs are usually extracted from sub-Coulomb transfer reactions which are extremely sensitive to the energy of the interaction and have a small cross section, the extracted partial widths often have significant uncertainties. A good example of the difficulty of measuring ANCs is given by the factor 3 change in the ANC for ^{13}C+α between the works of Johnson *et al.* [8] and Avila *et al.* [9], the latter study demonstrating that the former had an uncorrected problem due to build-up on the target reducing the effective beam energy. These extremely experimentally challenging studies are vital for nuclear astrophysics but require considerable expertise.

Transfer reactions typically only provide information on one part of a resonance reaction. However, since the cross section is typically dominated by the slowest step (i.e. the smallest partial width), the information obtained from the transfer reaction may not entirely specify the reaction rate. An excellent example of this was the ^{25}Mg$(d,p)^{26}$Mg study of Chen *et al.* [11] and its comparison to ^{25}Mg+n measurements [12, 13]. Since the γ-ray partial width is generally smaller for most of the relevant states, the ^{25}Mg$(n,\gamma)^{26}$Mg cross section is governed by the γ-ray partial widths. Neutron widths can be determined from (d,p) reactions but these may not tell the whole story. Coincidence measurements of the $(d,p\gamma)$ reaction may be necessary. For many reactions, especially charged-particle reactions, the particle partial width is the smallest and therefore rate-limiting width which means that the reaction rate can be totally specified from the transfer reaction.

4 Summary

Transfer reactions remain one of the most powerful tools for determining thermonuclear reaction rates for nuclear astrophysics, having been a workhorse of the field for decades. Transfer reactions can be used to locate astrophysically important states and measure their properties so that astrophysical reaction rates may be calculated. As with every experimental technique, the devil is in the details. In this paper, the broad background of transfer reactions has been laid out along with some potential pitfalls and problems.

References

[1] C. Iliadis, Nuclear Physics of Stars, Wiley-VCH, Weinheim (2007)
[2] F. Hammache and N de Séréville, Frontiers in Physics **8** 602920 (2021)
[3] I. Thompson and F. Nunes, Nuclear Reactions for Astrophysics, Cambridge University Press, Cambridge (2009).
[4] H. Krauss, TEDCA, unpublished computer code (1992)
[5] P. Adsley *et al.*, Journal of Instrumentation **12**, T02004 (2017).
[6] R. Neveling *et al.*, Nuclear Instruments and Methods in Physics A **654**, 29-39 (2011).
[7] S. D. Binda, MSc Dissertation, University of the Witwatersrand, Johannesburg, South Africa, (2022); S. D. Binda *et al.*, paper in preparation.
[8] E. D. Johnson *et al.*, Physical Review Letters **97**, 192701 (2006).
[9] M. L. Avila *et al.*, Physical Review C **90**, 024327 (2014).
[10] D.S. Harrouz *et al.*, Physical Review C **105**, 015805 (2022).
[11] Y. Chen *et al.* Physical Review C **103** 035809 (2021)
[12] C. Massimi *et al.* Physics Letters B **768** 1 (2017)
[13] C. Massimi *et al.* Physical Review C **85** 044615 (2012)

Nuclear astrophysics studies with γ-ray beams: What do we expect to learn from them?

Dimiter L. Balabanski[1*]

[1]Extreme Light Infrastructure – Nuclear Physics (ELI-NP), Horia Hulubei National Institute for R&D in Physics and Nuclear Engineering, 077125 Bucharest – Magurele, Romania

Abstract. An overview of the main directions of present-day studies with quasimonochromatic γ beams is discussed with an emphasis on the research opportunities which will be offered at the Extreme Light Infrastructure Nuclear Physics (ELI-NP) facility at Magurele near Bucharest in Romania. Experiments with γ beams at the extremes of high temperatures are outlined, with an emphasis on prospective studies related to nuclear astrophysics and astroparticle physics. Some of the experimental setups for nuclear structure, reaction, and astrophysics studies, which are available at ELI-NP, are described.

1 Introduction

In a recent review [1], the status and perspectives of photonuclear research were discussed, covering both fundamental science and applications. The development of narrow-bandwidth γ beams with sufficient brilliance, based on the laser Compton backscattering (LCB) technique [2,3], made possible detailed studies of lowest-lying excitation modes in atomic nuclei, *i.e.*, $E1$, $M1$, and $E2$ excitations. In experiments with LCB γ beams [4,5], it was demonstrated that nuclear dipole response can be studied at a new level of precision and sensitivity. Thus, experiments with LCB γ beams became the ideal tool for studies of large-amplitude collective excitations in atomic nuclei, such as the isovector giant dipole resonance (IVGDR) or the pigmy dipole (PDR) resonances.

This paper focuses on prospective studies related to nuclear astrophysics and astroparticle physics at ELI-NP. The γ-beam physics program at ELI-NP has been reviewed and updated recently [6-8]. Measurements of photoneutron reaction cross sections in light nuclei address astroparticle problems such as *e.g.* studies related to ultra-high energy cosmic rays (UHECR) [9,10]. The objectives of the PANDORA project, which aims at UHECR interdisciplinary studies, are briefly summarized and the proposed photonuclear experiments are discussed. Photon-induced charged-particle reactions are related to nuclear astrophysics. The perspectives for such studies together with the instruments, which are constructed at ELI-NP, are discussed. An important development in the field is the utilization of active targets. The program for active-target experiments and the detectors, which are available or are under construction at ELI-NP, are presented. The diverse opportunities for nuclear structure

* Corresponding author: dimiter.balabanski@eli-np.ro

research with vortex photons are discussed, too. Examples of possible experiments with high orbital angular momentum (OAM) γ rays are highlighted.

2 Nuclear structure studies at high nuclear temperatures

Traditionally, photonuclear excitations were used for studies at the extreme of high nuclear temperatures. The interaction of photons with atomic nuclei is a very selective process. They induce dipole and, to a much lesser extent, quadrupole excitations due to the angular momentum selection rules. This allows investigation of dipole and quadrupole excitations over a large range of energies. Studies below and above the particle evaporation threshold are performed. In NRF experiments, it is possible to tune the quasimonochromatic γ beam such that a discrete nuclear level is excited and thus to study its decay. For details of the NRF formalism and methodology see Ref. [1]. Above the particle threshold, states in the continuum are excited. The deexcitation goes either through the emission of gamma rays, evaporation of neutrons or charged particles, or fission.

Two detector arrays were constructed at ELI-NP for such studies. The ELI-NP Array of Ge DEtectors (ELIADE) [11,12] is a dedicated detector system for NFR experiments with quasimonochromatic γ beams. It is composed of eight segmented HPGe Clover detectors, placed in the horizontal and vertical planes with respect to the beam axis, organized in two rings, at 90^0 and 135^0, correspondingly. A recent photograph of the array is presented in Figure 1a. The overall efficiency of the array is about 6%. This allows measurements of γ-ray energies, intensities, angular distributions and polarization asymmetries, as well as γγ coincidences. Thus, it will be possible to extract from the measured spectra the γ-ray branching ratios, the γ-ray multipolarities and mixing ratios, the level energies, the spins, parities, and the *K* quantum numbers of the levels, their partial and total decay widths, the transition and excitation strength, and the integrated (γ,γ′) cross-section. The performance of the ELIADE detectors was tested in-beam and the results meet the requirements for high-precision NRF experiments [13].

Fig. 1. (a) The ELIADE detector in the E8 experimental hall of ELI-NP, and (b) the ELIGANT-GN detector in the E9 experimental hall of ELI-NP. Photograph courtesy of Dmitry Testov (ELIADE) and Pär-Anders Söderström (ELIGANT-GN).

The family of ELI-NP Gamma Above the Neutron Threshold (ELIGANT) instruments will be used to study (γ,γ′), (γ,n), (γ,γ′n) and (γ,νn), ν = 1,2,3 reactions. Thus, photoneutron cross sections will be measured with the ELIGANT - Thermal Neutron (ELIGANT-TN) setup [14], while the ELIGANT - Gamma and Neutron (ELIGANT-GN) detector array [14,15] will be used for measurements of γ rays and neutrons emitted during the de-excitation PDR or GDR states above the particle evaporation threshold. The array, which is displayed in Figure 1b, is made of two detector assemblies, an inner one, which is placed downwards at backward angles with respect to the beam at 30 cm from the target, and an outer one, placed in the upper hemisphere at 1.5 m from the target. The inner array is used for γ-ray detection and consists of 19 $CeBr_3$ and 15 $LaBr_3$:Ce scintillation detectors. The outer array consists of 37 EJ-309 liquid scintillation detectors and 25 GS20 ^{6}Li glass detectors, and is used for the detection of neutrons. ELIGANT-GN was commissioned with radioactive sources [16] and the performance was found to correspond to the design specifications [14], as well as to the response simulations [15].

2.1 Studies of photonuclear reactions at ELI-NP

Photodisintegration is one of the basic research techniques, which will be used at ELI-NP. Experimental setups for measurements of photoneutron and charged-particle reactions were built and are ready to take the first beams, which are expected to be delivered in 2023. The experimental techniques, the observables, and the quantities, which are deduced in these experiments, were reviewed in Ref. [1]. The expected increase of the spectral density of the γ beams at the VEGA facility ELI-NP with about two orders of magnitude with respect to the current state of the art will bring the experimental techniques to a new level of sensitivity. The experimental program at the facility, which has been updated recently by Tanaka et al. [8], includes studies related to nuclear astrophysics and precise measurements of photoneutron cross sections.

Photonuclear reactions are one of the mechanisms involved in nucleosynthesis. Photons from the high-energy tail of the Plank distribution can photodisintegrate nuclei. This is demonstrated best in the synthesis of some proton-rich nuclei heavier than iron via the astrophysical *p*-process, where (γ,n), (γ,p) and (γ,α) reactions are one of the driving forces [17]. Another example, which underlines the importance of photonuclear reaction studies in astrophysics is related to studies of time-reverse reactions, such as, *e.g.*, ^{16}O(γ,α)^{12}C reaction. In this experiment, due to the low reaction cross section close to the Gamow window, a time projection chamber (TPC), with optical readout, *i.e.*, an active target, was used [18]. The experiment was recently repeated at the HIγS free-electron laser γ-ray facility at Duke University [19], using the newly developed Warsaw TPC with an electronic readout.

In active-target experiments, the target, which is a gas, serves also as a detector medium. Charged particles coming from photonuclear reactions produce drifting electrons in the active volume of the TPC. At ELI-NP, two active-target experimental setups based on the time projection chamber (TPC) technology will be available, the mini-TPC and the ELITPC [20]. The mini-TPC is available for experiments and the ELITPC is still under implementation. Both of them use an electronic readout based on the GEM detector technology [21]. The difference between the two detectors is their active volume and number of readout channels. The mini-TPC utilizes a 256-channel readout in the horizontal plane, perpendicular to the gradient of the applied voltage which forces electrons to drift. It consists of three layers of electrodes, referred to as *u-v-w* readout. They are oriented at 120^0 to each other and placed on a multilayer PCB. The ELITPC, which is a replica of the Warsaw TPC, will utilize a 1024-channel readout. The signals from the readout are processed by GET (General Electronics for TPCs) boards [22].

In addition, a 4π array of Si strip detectors, the ELI-NP Si strip detector array (ELISSA) is available for charged-particle experiments at ELI-NP [20]. It consists of 36 X3-type double-sided Si strip detectors (DSSSD), which form a barrel, and eight QQQ3-type DSSSDs, which form endcaps. The barrel is built and tested and the endcaps are still to be implemented. The total angular coverage of the instrument will be 80%, and it will detect protons of E_p = 100 keV – 10 MeV and α particles of $E_α$ = 100 keV – 30 MeV, *e.g.* with outstanding energy and position resolution. An experimental setup, combining the X3 ELISSA DSSSD detectors in a barrel configuration with the LHASA YY2 DSSSD detectors in a lamp-shape configuration, was used for the measurement of the ^{18}F(p,α) reaction at the 3 MV tandem accelerator at IFIN-HH. The experimental setup is shown in Figure 2.

Fig. 2. (a) Schematic drawing of the ELISSA+LHASA experimental setup, and (b) the ELISSA+LHASA detector array in a reaction chamber at the 3 MV tandem accelerator of IFIN-HH. Photograph courtesy of Teodora Madgearu.

A proof-of-principle experiment was carried out at the HIγS facility, at Duke University. There the photodisintegration of ^{7}Li was studied [23], which is of importance for the understanding of the Big Bang nucleosynthesis. For this purpose, quasimonoenergetic photons in the energy range from 4.4 to 10 MeV were used, and tritons and α particles were detected in coincidence by segmented Si detectors. These studies will be continued at ELI-NP at lower energies and with better resolution in the resonance regions. Further experiments at HIγS, which will be carried out in early 2023, will include (γ,p) and (γ,α) reaction studies of the key *p*-process nuclei ^{102}Pd and ^{112}Sn.

2.2 Photonuclear reactions of light nuclei for studies of ultra-high energy cosmic rays

The PANDORA project aims to systematically measure the photonuclear response of nuclei with A ≤ 56. The experiments will be carried out at RCNP, at the University of Osaka, and iThemba Labs, Cape Town with virtual photons in (p,p′) reactions at zero degrees, and at ELI-NP with real photons.

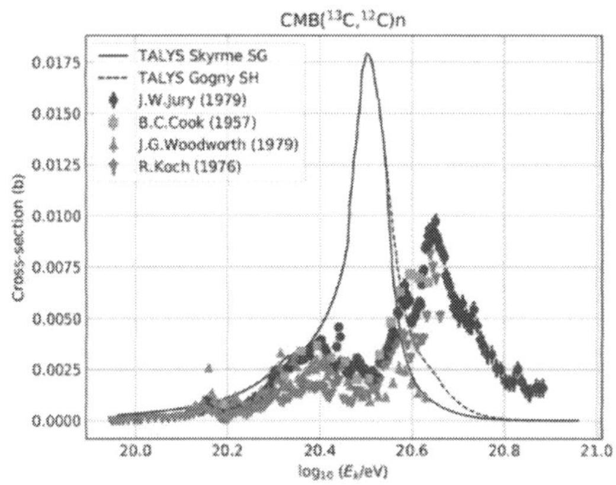

Fig. 3. Comparison of the experimental data for the cross-section of the ^{13}C(γ,n) reaction and theoretical calculations within the TALYS computer code using Skyrme and Gogny parametrizations of the level densities.

The project is of particular interest related to studies of the propagation of ultra-high energy cosmic rays (UHECR). The primary disintegration of the UHECRs in extra-galactic space is through the interaction of the particle species with the strongly Doppler-shifted cosmic microwave background leading to photonuclear reactions mainly in the GDR region. The project is carried out in synergy with the UHECR theory groups. It is intending on providing reliable photo-nuclear cross-section inputs for a large set of nuclei, which would significantly improve our understanding of the UHECR propagation and generation model. On the nuclear physics side, there is a strong need for good quality photonuclear reaction data for most of the light nuclei of astrophysical interest below iron. The situation is further complicated by the non-availability of reliable theoretical means for required photo-nuclear cross sections since the prediction of mean-field theories and statistical models do not agree well with measured values in this region, as illustrated in Figure 3 for ^{13}C. These measurements would lead to new insights into the effect of clustering, shell structure, and NN correlations on the photonuclear response of these nuclei.

The existing data for the $^{13}C(\gamma,n)$ reaction in inverse kinematics, as a function of ^{13}C kinetic energy where the cosmic microwave background serves as a target. The three facilities together will cover the full energy region of the shown data. The highest energy cosmic rays observed so far reach an energy of up to approximately $10^{20.5}$ eV. The data [24-27] demonstrate a discrepancy between themselves, but they also are off theoretical calculations which were done with the TALYS computer code [28] using Skyrme and Gogny parametrizations for the level densities.

2.3 Nuclear astrophysics with OAM photons

Astrophysics models consider also the creation of OAM photons in the Universe in the presence of strong magnetic fields [29-31]. This opens the need for revisiting a wide range of nuclear photonics problems, such as photodisintegration reactions. So far, all photodisintegration reactions were considered to be induced by normal photons, having spin $1\hbar$. Since there is not a single facility in the world where vortex γ rays can be generated, ongoing research is focused on theory. The modification of the photonuclear reaction cross sections in experiments with twisted photons, having angular momentum $\geq 2\hbar$, were evaluated [32], as well as the generalization of the NRF formalism for higher angular momenta of the incident photons and possible experimental applications are studied.

3 Conclusions and outlook

Photonuclear reactions attract serious interest in the understanding of a number of problems related to nuclear astrophysics and astroparticle physics. Several examples which discuss demonstrate the potential and the need for such studies. These include the understanding of the propagation of ultra-high energy cosmic rays, which requires more precise measurements of the photodisintegration reaction cross sections in light nuclei, as well as further development of the theoretical models. Other examples include studies of photonuclear reactions in active-target experiments, measurements of γ strength functions and nuclear level densities, *etc.* Such measurements will become possible with the implementation of the VEGA facility at ELI-NP. The needed equipment for the performance of these measurements is available at ELI-NP.

This work was supported by a grant from the Romanian Ministry of Research, Innovation and Digitization, CNCS – UEFISCDI, project number PN-III-P4-PCE-2021-0595, within PNCDI III program.

References

1. A. Zilges, D.L. Balabanski, N. Pietralla, J. Isaak, Prog. Part. Nucl. Phys. **122**, 103903 (2022)
2. R.H. Milburn, Phys. Rev. Lett. **10**, 75 (1963)
3. F.R. Arutyunian, V.A. Tumanian, Phys. Lett. **4**, 176 (1963)
4. H. Ohgaki, T. Noguchi, S. Sugiyama, T. Yamazaki, T. Mikado, M. Chiwaki, K. Yamada, R. Suzuki, N. Sei, Nucl. Instrum. Meth. Phys. Res. A **353**, 384 (1994)
5. N. Pietralla, Z. Berant, V.N. Litvinenko, S. Hartman, F.F. Mikhailov, I.V. Pinayev, G. Swift, M.W. Ahmed, J.H. Kelley, S.O. Nelson, R. Prior, K. Sabourov, A.P. Tonchev, H.R. Weller, Phys. Rev. Lett. **88**, 012502 (2002)
6. D.L. Balabanski, R. Popescu, D. Stutman, K. Tanaka, O. Tesileanu, C.A. Ur, D. Ursescu, N.V. Zamfir, Europhys. Lett. **117**, 28001 (2017)
7. S. Gales, K.A. Tanaka, D.L. Balabanski, F. Negoita, D. Stutman, O. Tesileanu, C.A. Ur, D. Ursescu, S. Ataman, M.O. Cernaianu, L. D'Alessi, I. Dancus, B. Diaconescu, N. Djourelov, D. Filipescu, P. Ghenuche, D.G. Ghita, C. Matei, K. Seto, M. Zeng, N.V. Zamfir, Rep. Prog. Phys. **81**, 094301 (2018)
8. K.A. Tanaka, K.M. Spohr, D.L. Balabanski, S. Balascuta, L. Capponi, M.O. Cernaianu, M. Cuciuc, A. Cucoanes, I. Dancus, A. Dhal, B. Diaconescu, D. Doria, P. Ghenuche, D.G. Ghita, S. Kisyov, V. Nastasa, J.F. Ong, F. Rotaru, D. Sangwan, P.-A. Söderström, D. Stutman, G. Suliman, O. Tesileanu, N. Tsoneva, L. Tudor, C.A. Ur, D. Ursescu, N.V. Zamfir, Matter Radiat. Extremes **5**, 024402 (2020)
9. E. Khan, S. Goriely, D. Allard, E. Parizot, T. Suomijarvi, A.J. Koning, S. Hilaire, M.C. Duijvestijn, Astropart. Phys. **23**, 191 (2005)
10. R. Alves Batista, D. Boncioli, A. di Matteo, A. van Vliet, D. Walz, J. Cosmol. Astropart. Phys. 10, 063 (2015)
11. C.A. Ur, A. Zilges, N. Pietralla, J. Beller, B. Boisdeffrw, M. Cernaianu, V. Derya, B. Löbner, C. Matei, G. Pascovici, C. Petcu, C. Romig, D. Savran, G. Suliman, E. Udup, V. Verner, Rom. Rep. Phys. **68**, S438 (2016)
12. P.-A. Söderström, G. Suliman, C.A. Ur, D. Balabanski, T. Beck, L. Capponi, A. Dhal, V. Iancu, S. Ilie, M. Iovea, A. Kusoglu, C. Petcu, N. Pietralla, G.V. Turturica, E. Udup, J. Wilhelmy, A. Zilges, Acta Phys. Pol. B **50**, 329 (2019)
13. L. Capponi, A. Kuşoğlu, P.-A. Söderström, D.L. Balabanski, G.V. Turturică, G. Bocchi, S. Chesnevskaya, A. Dhal, D. Dinescu, N. Djourelov, Y. Niu, A. Oprisa, A. Pappalardo, G. Suliman, C.A. Ur, J. Instrum. **16**, T12001 (2021)
14. F. Camera, H. Utsunomiya, D. Varlamov, D. Filipescu, V. Baran, A. Bracco, G. Colò, I. Gheorghe, T. Glodariu, C. Matei, O. Wieland, Rom. Rep. Phys. **68**, S539 (2016)
15. M. Krzysiek, F. Camera, D.M. Filipescu, H. Utsunomiya, G. Colò, I. Gheorghe, Y. Niu, Nucl. Instrum. Meth. Phys. Res. A **916**, 257 (2019)
16. Pär-Anders Söderström, Esra Açıksöz, Dimiter Balabanski, Franco Camera, Luigi Capponi, Gheorghe Ciocan, Mihai Cuciuc, Dan Filipescu, Ioana Gheorghe, Tudor Glodariu, Jasmeet Kaur, Matteus Krzysiek, Catalin Matei, Traian Roman, Adrian Rotaru, Andreea Serban, Alexandru State, Hiroaki Utsunomiya, Vlad Vasilca, Nucl. Instrum. Meth. Phys. Res. A **1027**, 166171 (2022)
17. M. Arnould, S. Goriely, Phys. Rep. **384**, 1 (2003)
18. M. Gai, D. Schweitzer, S.R. Stern, A.H. Young, R. Smith, M. Cwiok, J.S. Bihalowicz, H. Czyrkowski, R. Dabrowski, W. Dominik, A. Fijalkowska, Z. Janas, L. Janiak, A.

Korgul, T. Matulewicz, C. Mazzocchi, M. Pfuetzner, M. Zaremba, D. Balabanski, I. Gheorge, C. Matei, O. Tesilenau, V.N. Zamfir, M.W. Ahmed, S.S. Henshaw, C.R. Howell, J.M. Mueller, L.S. Myers, S. Stave, C. Sun, H.R. Weller, Y.K. Wu, A. Breskin, V. Dangendorf, K. Tittelmeier, M. Freer, Nucl. Instrum. Meth. Phys. Res. A **954**, 161779 (2020)

19. V.N. Litvinenko, B. Burnham, M. Emamian, N. Hower, J.M.J. Madey, P. Morcombe, P.G. O'Shea, S.H. Park, R. Sachtschale, K.D. Straub, G. Swift, P. Wang, Y. Wu, R.S. Canon, C.R. Howell, N.R. Roberson, E.C. Schreiber, M. Spraker, W. Tornow, H.R. Weller, I.V. Pinayev, N.G. Gavrilov, M.G. Fedotov, G.N. Kulipanov, G.Y. Kurkin, S.F. Mikhailov, V.M. Popik, A.N. Skrinsky, N.A. Vinokurov, B.E. Norum, A. Lumpkin, B. Yang, Phys. Rev. Lett. **78**, 4569 (1997)
20. O. Tesileanu, M. Gai, M. Anzalone, C. Balan, J.S. Bihalowicz, M. Cwiok, W. Dominik, S. Gales, D. G. Ghita, Z. Janos, D.P. Kendellen, M. La Cognata, C. Matei, K.Mikszuta, C. Petcu, M. Pfützner, T. Matulewicz, C. Mazzocchi, and C. Spitaleri, Rom. Rep. Phys. **68**, S699 (2016)
21. F. Sauli, Nucl. Instrum. Meth. Phys. Res. A **386**, 531 (1997); *ibid* **805**, 2 (2016)
22. E.C. Pollacco, G.F. Grinyer, F. Abu-Nimeh, T. Ahn, S. Anvar, A. Arokiaraj, Y. Ayyad, H. Baba, M. Babo, P. Baron, D. Bazin, S. Beceiro-Novo, C. Belkhiria, M. Blaizot, B. Blank, J. Bradt, G. Cardella, L. Carpenter, S. Ceruti, E. De Filippo, E. Delagnes, S. De Luca, H. De Witte, F. Druillole, B. Duclos, F. Favela, A. Fritsch, J. Giovinazzo, C. Gueye, T. Isobe, P. Hellmuth, C. Huss, B. Lachacinski, A.T. Laffoley. G. Lebertre, L. Legeard, W.G. Lynch, T. Marchi, L. Martina, C. Maugeais, W. Mittig, L. Nalpas, E.V. Pagano, J. Pancin, O. Poleshchuk, J.L. Pedroza, J. Pibernat, S. Primault, R. Raabe, B. Raine, A. Rebii, M. Renaud, T. Roger, P. Roussel-Chomaz, P. Russotto, G. Saccà, F. Saillant, P. Sizun, D. Suzuki, J.A. Swartz, A. Tizon, A. Trifiró, N. Usher, G. Wittwer, J.C. Yang, Nucl. Instrum. Meth. Phys. Res. A **887**, 81 (2018)
23. M. Munch, C. Matei, S.D. Pain, M. T. Febbraro, K. A. Chipps, H.J. Karwowski, C.Aa. Diget, A. Pappalardo, S. Chesnevskaya, G.L. Guardo, D. Walter, D.L. Balabanski, F.D. Becchetti, C.R. Brune, K.Y. Chae, J. Frost-Schenk, M.J. Kim, M.S. Kwag, M. La Cognata, D. Lattuada, R.G. Pizzone, G.G. Rapisarda, G.V. Turturica, C.A. Ur, Y. Xu, Phys. Rev. C **101**, 055801 (2020)
24. J.W. Jury, B.L. Berman, D.D. Faul, P. Meyer, K.G. McNeill, J.G. Woodworth, Phys. Rev. C **19**, 1684 (1979)
25. J.G. Woodworth, K.G. McNeill, J.W. Jury, P.D. Georgopulos, R.G. Johnson, Nucl. Phys. A **327**, 53 (1979)
26. R. Koch, H.H. Thies, Nucl. Phys. A **272**, 296 (1976)
27. Barnett C. Cook, Phys. Rev. **106**, 300 (1957)
28. A.J. Koning and D. Rochman, Nucl. Data Sheets **113**, 2841 (2012)
29. Martin Harwit, Astrophys. J. **597**, 1266 (2003)
30. Fabrizio Tamburini, Bo Thidé, Gabriel Molina-Terriza, Gabriele Anzolin, Nature Phys. **7**, 195 (2011)
31. Tomoyuki Maruyama, Takehito Hayakawa, Toshitaka Kajino, Myung-Ki Cheoun, Phys. Lett. B **826**, 136779 (2022)
32. Yi Xu, D.L. Balabanski, V. Baran, C. Matei, Phys. Rev. C, in preparation

Big Bang nucleosynthesis as a probe of new physics

Carlos A. Bertulani[1,*], *Francis W.* Hall[1,**], and *Benjamin I.* Santoyo[1,***]

[1]Department of Physics and Astronomy, Texas A&M University-Commerce, Commerce, TX, United States

Abstract. The Big Bang Nucleosynthesis (BBN) model is a cornerstone for the understanding of the evolution of the early universe, making seminal predictions that are in outstanding agreement with the present observation of light element abundances in the universe. Perhaps, the only remaining issue to be solved by theory is the so-called "lithium abundance problem". Dedicated experimental efforts to measure the relevant nuclear cross sections used as input of the model have lead to an increased level of accuracy in the prediction of the light element primordial abundances. The rise of indirect experimental techniques during the preceding few decades has permitted the access of reaction information beyond the limitations of direct measurements. New theoretical developments have also opened a fertile ground for tests of physics beyond the standard model of atomic, nuclear, statistics, and particle physics. We review the latest contributions of our group for possible solutions of the lithium problem.

1 Introduction

From about 10 seconds to 20 minutes after the Big Bang the universe endured the production of light elements in a process known as Big Bang Nucleosynthesis (BBN) [1–14]. The temperature during this epoch decreased from about 10^9 K to 10^8 K. BBN and its predictions depend strongly on few quantities, namely, (a) the ratio of baryon-to-photon number densities, $\eta = n_b/n_\gamma$, assumed to be uniform in space and time during BBN, (b) the number of neutrinos families N_ν, and (c) the neutron half-life τ_n. The Cosmic Microwave Background (CMB) measurement obtained with the Planck satellite yields $\eta = (6.12 \pm 0.06) \times 10^{-10}$ [15]. Precise experiments at the CERN Large Electron–Positron Collider (LEP) have deduced the number of neutrino families [16] as $N_\nu = 2.9840 \pm 0.0082$, while neutron lifetime measurements have obtained $\tau_n \simeq 877.75 \pm 28$ s [17]. All these values are in excellent agreement with the BBN model and its predictions of the abundance of light elements. In particular, the BBN model predicts $\eta = (6.123 \pm 0.039) \times 10^{-10}$, compatible with the most recent measurements of the deuterium primordial abundance [18]. Nowadays, it is possible to say that the BBN is a parameter-free model, calculated using rather simple computer programs and yielding consistent predictions for primordial abundances with inputs of experimental nuclear cross sections.

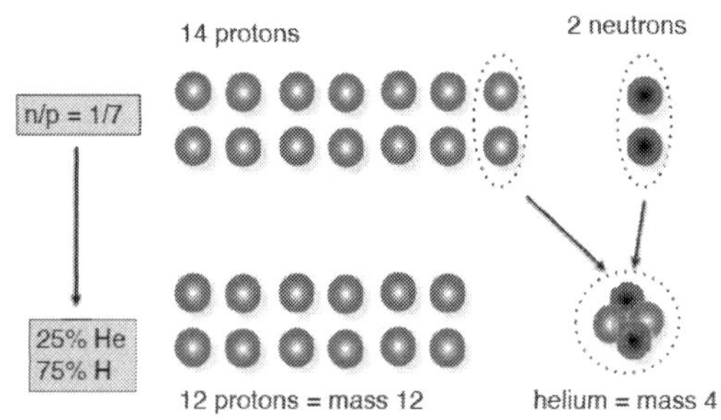

Figure 1: *Left:* The observed nearly 25% of helium and 75% of hydrogen in the universe can be understood if the neutron to proton ratio during the Big Bang nucleosynthesis was equal to n/p = 1/7.

During the BBN, the light elements D, ^{3}He, ^{4}He, and ^{7}Li were produced and their abundances have been observed in particular astrophysical environments thought to be remnants of the BBN epoch. The BBN predictions depend on the magnitude of the nuclear cross sections and the nuclear reaction network in a medium where densities and temperatures evolve within the model. One solves the Friedmann's equations for the expansion rate of the universe,

$$H^2(t) = \left(\frac{da/dt}{a}\right)^2 = \frac{8\pi G}{3}\rho - \frac{k}{a^2} + \frac{\Lambda}{3}, \qquad (1)$$

where $H(t)$ is the evolving Hubble constant, a is a universal scale factor for distances, k is the "curvature" and Λ is the cosmological constant. The universe is assumed to be homogeneous and isotropic (cosmological principle). The Friedmann-Lemaitre-Robertson-Walker (FLRW) metric is adopted to describe distances between neighboring points

*e-mail: carlos.bertulani@tamuc.edu
**e-mail: fhall2@leomail.tamuc.edu
***e-mail: ben@santoyo.us

in the space-time,

$$ds^2 = dt^2 - a^2(t)\left[\frac{dr^2}{1-kr^2} + r^2\left(d\theta^2 + \sin\theta d\phi^2\right)\right]. \quad (2)$$

Additionally, the first law of thermodynamics invokes a relation between density ρ and pressure p in the form

$$\frac{d\rho}{dt} = -3H(\rho + p). \quad (3)$$

Moreover, charge-equilibrium implies that

$$n_b \sum_j Z_j X_j = n_{e^+} - n_{e^-} = \Phi\left(\frac{m_e}{T}, \mu_e\right), \quad (4)$$

where $X_i = n_i m_i/\rho$ is the mass fraction of the particle species i, with mass m_i and number density n_i. Nuclear abundances Y_i are defined so that $X_i = A_i Y_i$ is the mass fraction in the environment for a nuclide with charge Z_i and mass number A_i, with the normalization $\sum_i A_i Y_i = 1$. The function Φ incorporates the densities of electrons, positrons and neutrinos. Weak-decays also change the relative numbers of e^+ and e^- through $n \leftrightarrow p$ rates.

All nuclear reaction rates include at most four different nuclides and are of the form

$$N_i(^{A_i}Z_i) + N_j(^{A_j}Z_j) \leftrightarrow N_k(^{A_k}Z_k) + N_l(^{A_l}Z_l), \quad (5)$$

where N_m is the number of nuclide m, with $A_i \geq A_j$ and $A_l \geq A_k$. The abundance changes are given in terms of dY_i/dt which is computed as

$$\frac{dY_i}{dt} = \sum_{j,k,l} N_i\left(-\frac{Y_i^{N_i} Y_j^{N_j}}{N_i! N_j!}[ij]_k + \frac{Y_l^{N_l} Y_k^{N_k}}{N_l! N_k!}[lk]_j\right), \quad (6)$$

where $[ij]_k$ is the forward reaction rate and $[lk]_j$ is the reverse rate and abundance changes are summed over all forward and reverse reactions involving nuclide i.

The reaction rates in a stellar environment with a temperature T is given by a Maxwell-Boltzmann average over the relative velocity distribution of the pair $[ij]$,

$$[ij]_k = \langle\sigma v\rangle_{[ij]_k} = \sqrt{\frac{8}{\pi\mu(kT)^3}} \int_0^\infty \sigma(E) E \exp\left(-\frac{E}{kT}\right) dE, \quad (7)$$

where μ is the reduced mass of $[ij]$, N_A is Avogadro's number, k is the Boltzmann constant, and the reaction cross section and its energy dependence is denoted by $\sigma(E)$.

Using $k = 0$ and $\Lambda = 0$ for the epoch right after the baryogenensis, the standard BBN model described above leads to the formation of deuterons about $\sim$ 3 minutes after the Big Bang. The p(n,γ)d reaction flow for deuteron formation is strongly dependent on the value of η. Deuterons were immediately destroyed with the creation of ^{3}He nuclei by means of d(p,γ)^{3}He and d(d,n)^{3}He reactions and the formation of tritium with the d(d,p)t reaction. Subsequently, ^{4}He were formed via the ^{3}He(d,p)^{4}He and t(d,n)^{4}He reactions. This chain of reactions yields a universe with about 75% of hydrogen and 25% of helium. As the universe expanded and cooled down, nucleosynthesis continued at a smaller rate, leading to very small amounts of ^{7}Li and ^{6}Li. Heavier elements such as carbon, nitrogen or oxygen were created in very tiny amounts, of the order of < 10^{16} relative abundance [19].

The prediction of 75% of hydrogen and 25% of helium mass fraction in the universe is one of the major accomplishments of the Big Bang model [8]. It solely relies on the the neutron-to-proton ratio being equal to n/p = 1/7 at the BBN epoch, as schematically shown in Fig. 1. The n/p = 1/7 value is obtained by solving the Big Bang standard model (see Figure 1 of Ref. [8]).

The most relevant BBN reactions are listed in Table 1. The predicted mass and number fractions of light elements produced during the BBN are shown in Fig. 2 as a function of time. One notices that elements up to ^{7}Be and ^{7}Li are predicted in reasonable amounts. But ^{7}Be decays by electron capture in about 53 days. Therefore, all ^{7}Be produced during the BBN is observed today as ^{7}Li.

n $\leftrightarrow$ p	p(n,γ)d	d(p,γ)^{3}He	d(d, p)t
d(d, n)^{3}He	^{3}He(n, p)t	t(d, n)^{4}He	^{3}He(d, p)^{4}He
^{3}He(α, γ)^{7}Be	t(α, γ)^{7}Li	^{7}Be(n, p)^{7}Li	^{7}Li(p,α)^{4}He

Table 1: Most relevant BBN reactions.

BBN predicts a ^{7}Li/H of abundance of $\sim 10^{-10}$ and a ^{6}Li/H abundance of $\sim 10^{-14}$. After the BBN era ($\gtrsim$ 20 min), the universe cooled down and no more elements were formed until the first stars were born, about 100 million years later. After the BBN and substantial formation of stars, ^{6}Li and ^{7}Li could be produced in spallation processes by cosmic rays formed by star ejecta. ^{7}Li can also be synthesized in novae or during AGB stars pulsations. Astronomical observation suggest that the ^{7}Li abundance does not depend on "metallicity" (the content of elements heavier than ^{4}He) in metal-poor stars. These stars have small Fe/H abundances and are observed in the galaxy halo. The ^{7}Li abundance in such stars is nearly constant, a phenomenon known as the "Spite plateau" [20].

Low metallicity stars with masses $M < M_\odot$ and life expectancies greater than the age of the universe have (in principle) no convective zone near their surfaces. Lithium present in their surfaces cannot be brought to deeper zones and be destroyed where temperatures exceed 1 GK. The depth of the convection zone decreases with the surface temperature, and that is the reason why astronomers select stars with T_{eff} > 6000 K (warm) for such observations (for more details, see Ref. [21]). Other observations of low-metallicity stars apparently contradict the inferences based on the Spite plateau [22, 23]. Using a variety of observations and laboratory data on nuclear reactions, and $\eta = (6.07 \pm 0.07) \times 10^{-10}$ [15], the BBN model predicts a ^{7}Li abundance of Li/H = $(4.16 - 5.34) \times 10^{-10}$ [25] whereas observations from metal-poor halo stars obtains Li/H = $(1.58 + 0.35 - 0.28) \times 10^{-10}$ [23, 24]. This value is about 3 times smaller than the one predicted by the BBN model and is the source of the so-called "lithium puzzle".

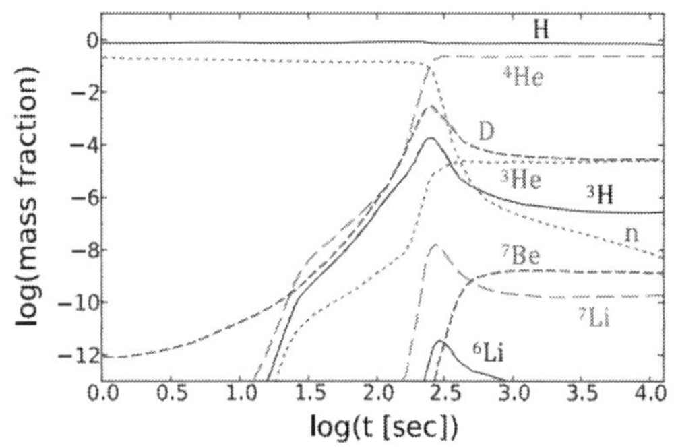

Figure 2: Mass (for ^{4}He) and number (for other elements) fractions of nuclei in the universe as a function of time.

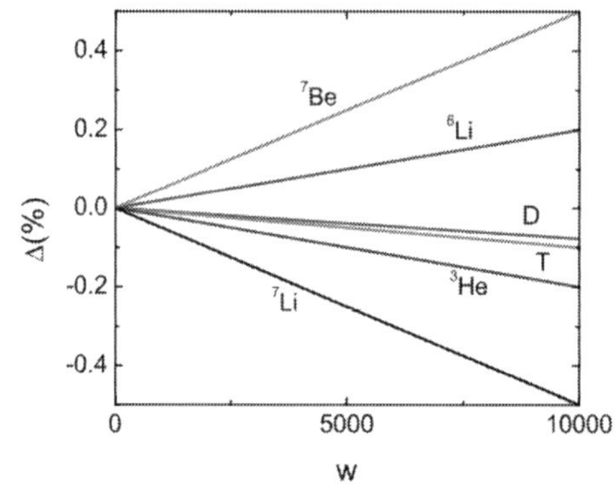

Figure 3: Variation (in percent) of the abundances of light nuclei produced in the BBN as a function of an artificial multiplicative factor w enhancing Salpeter's formulation of screening [34].

For some time, it was apparent that there was a second lithium puzzle, involving the BBN ^{6}Li abundance via the ^{2}H$(\alpha,\gamma)^6$Li reaction. ^{6}Li (and ^{7}Li) is also formed in cosmic ray spallation reactions in the interstellar medium out of which (halo) stars are born. The second lithium puzzle concerns the BBN predictions for the isotopic ratio ^{6}Li/^{7}Li $\sim 10^{-5}$ [25, 26] being smaller than the observations ^{6}Li/^{7}Li $\sim 5\times10^{-2}$ [27]. This puzzle is controversial due to the complexities of 3-dimension calculations with convection and non-local thermodynamical equilibrium in the photo-sphere of metal-poor stars. These complexities weaken the possibility of a second lithium problem for the ^{6}Li/^{7}Li isotopic ratio. Recent observations seem to indicate that the second cosmological lithium problem does not exist, based on the lack of evidence of ^{6}Li in Spite Plateau stars [28, 29].

2 Nuclear and atomic physics

As mentioned above, the most relevant BBN nuclear reactions are shown in Table 1. The reaction network leads a significant production of D, ^{3}H, ^{3}He, ^{4}He, ^{6}Li, ^{7}Li and ^{7}Be and to the production of small traces of carbon, nitrogen and oxygen, at the $10^{-15}-10^{-25}$ abundance yield. It is nearly useless to make predictions beyond such nuclei because they are overwhelmingly produced in stars although the inclusion of standard nuclear reactions networks, such as the famous CNO cycle have been carried out for clarity [24].

Our calculations presented in Fig 2 for the BBN abundances of H, D, ^{3}H, ^{3}He, ^{4}He, ^{6}Li, ^{7}Li and ^{7}Be as a function of time [10] have been performed with an extended version of a code originally due to Wagoner [30], reformulated by Kawano [31] and later modified to include the most recent updated nuclear reaction rates [32] and new physics aspects discussed in this short review. In Fig. 2 the ^{4}He mass fraction is represented by the dashed red curve, the deuterium abundance ratio D/H by the blue dashed curve, the ^{3}He /H abundance by the dotted green curve, the ^{3}H/H abundance by the solid black curve, the ^{7}Be/H abundance by a blue dashed curve, and the ^{7}Li/H abundance by a red dashed curve. New experimental measurements of nuclear reaction cross sections, including those reported with the Trojan Horse Method (THM) [25], have been incorporated in our compilation using R-matrix fits [32] .

2.1 Electron screening and clustering effects

One of the problems in deducing the reaction cross sections at the low astrophysical energies needed for the BBN is the cross section enhancement due to the electrons in the atomic targets in laboratory experiments. Also, within in the stellar plasma, free electrons are further responsible for enhancements of the reaction rates. The reaction rate enhancement in the plasma can be approximated rather well by the factor

$$f(E)=\frac{\sigma_s(E)}{\sigma_b(E)}, \tag{8}$$

where σ_s denotes the screened and σ_b the non-screened, or "bare" fusion cross section. Using the Debye-Hückel approximation the screened Coulomb potential of a nuclear species i can be written as [33]

$$V_i(r)=\frac{e^2Z_i}{r}\exp\left(-\frac{r}{R_D}\right). \tag{9}$$

The Debye radius entering this equation is related to the ion number density n_i and all environment charges j as

$$R_D=\frac{1}{\zeta}\left(\frac{kT}{4\pi e^2 n_i}\right)^{1/2}, \tag{10}$$

where

$$\zeta = \left[\sum_j X_j \frac{Z_j^2}{A_j} + \chi \sum_j X_j \frac{Z_j}{A_j} \right]^{1/2}. \quad (11)$$

Here, X_j denotes the mass fraction of particle j, T_6 the temperature in units of 10^6 K and χ is a correction due to effects of electron degeneracy. During the BBN, electrons and positrons are produced via the reaction $\gamma\gamma \rightarrow e^+e^-$. As the universe cooled down, the number density of electrons decreased strongly with T_6, from about 10^{32}/cm^3 at 10^{10} K to 10^{20}/cm^3 K at 3×10^8 K [34]. For comparison, the electron density in the core of the sun ($T = 15 \times 10^6$ k) is $n_e^{sun} \sim 10^{26}$/cm^3. But the baryon number density is much lower during the BBN than in the core of the sun and, due to charge neutrality, the number of excess electrons during the BBN is roughly equal to the number of protons. These excess electrons are the most relevant for the screening effect.

An attempt was reported in Ref. [34] to understand the effects of electron screening in the BBN production of light elements finding that these effects are negligible. A fudge factor w artificially reducing the Debye radius R_D was introduced to simulate any unusual enhancement of the electron screening in nuclear reactions during the BBN. The results for the modified light element abundances are shown in Fig. 3. The abundances are very resilient to a modification of the electron screening even if an extremely small Debye radius is assumed. Later, additional works have extended this study to include relativistic corrections [35, 36].

It became clear that electron screening cannot be responsible for the deficiency in the observed lithium abundance. Another publication studied the clustering in light nuclei and their influence in nuclear reactions at the BBN relevant energies. Clustering could perhaps explain part of the discrepancies of the experimental extraction of the electron screening enhancement factor $f(E)$ and the predictions of theory [37, 38].

2.2 Nuclear reaction rates

In Table 2 we compare our BBN calculations with observations. Y_p is a historical denomination of the ^{4}He mass fraction. Its value is taken from Ref. [39]. The deuterium abundance D/H = $(2.527 \pm 0.03) \times 10^{-5}$ adopted from Ref. [40, 41] is in accordance with the value $100\Omega_b h^2$ (BBN) = 2.225 ± 0.016 obtained from the observations of the cosmic microwave background [15]. Here, $\Omega_b h^2$ is the baryon density, where h is the reduced Hubble constant [13]. The ^{3}He/H abundance is extracted from Ref. [42]. Finally, the lithium abundance is adopted from Ref. [23]. It is in visible disagreement with astronomical observations (BBN/obs ~ 3).

Numerous attempts have been discussed in the literature as possible ways to explain the discrepancy between the observed and the calculated ^{7}Li abundance. A few attempts concentrated on the ^{7}Be abundance because ^{7}Be could be destroyed during the BBN. ^{7}Be decays in about

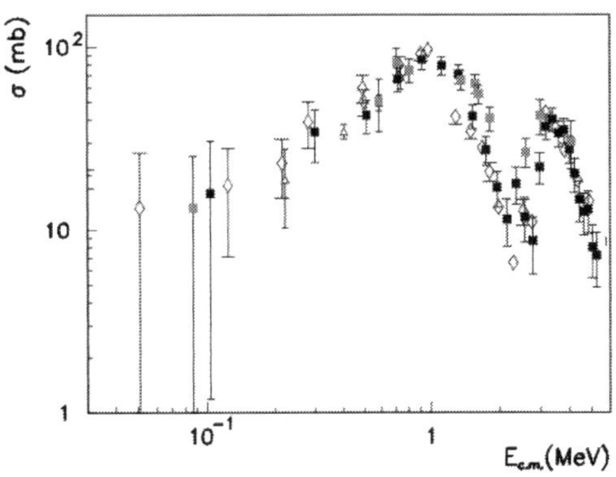

Figure 4: ^{7}Be(n,α) cross section obtained with the THM using the ^{7}Li(p,α)^{4}He mirror reaction (full red squares) and ^{3}He THM breakup data (full black squares) [47]. Other data compiled in Ref. [44] are displayed as empty blue diamonds. Additionally, data from Ref. [46] are displayed as empty blue triangles.

53.22 ± 0.06 days by electron capture to the ^{7}Li ground state and also to the first excited (0.477 MeV) of ^{7}Li. Thus, ^{7}Be created during the BBN will count towards the ^{7}Li primordial abundance. ^{7}Be could be destroyed by (n,p) and/or (n,α) reactions and explain the smaller lithium abundance. Refs. [43–47] have experimentally investigated this possibility. For example, Lamia et al. [47] measured the ^{7}Be(n,α) reaction cross section using the Trojan Home Method (THM) for the ^{7}Li(p,α)^{4}He mirror reaction and corrected it to account for Coulomb effects, as shown in Figure 4. The cross section for ^{7}Be(n,α) obtained with the THM is located within the Gamow window at the BBN temperatures and leads to a reaction rate smaller by a factor ~ 10 to the one adopted by Wagoner [30]. A BBN calculation using the new reaction rate results in a ^{7}Li/H abundance of 2.845×10^{-11} and a ^{7}Be/H abundance of 4.156×10^{-10}. This modification yields a BBN lithium/H abundance of 4.441×10^{-10}. No substantial modification of the BBN abundances prediction for other light nuclei is confirmed.

A theoretical study of the effect of ^{7}Be(α, γ) on ^{7}Be destruction was further carried out in Ref. [48], implying that the ^{7}Be abundance would only be changed appreciably if an unexpectedly strong resonance would exist very close to energy threshold for γ emission. At present, it appears that this resonant state has not been identified experimentally.

Another attempt to explain the lithium puzzle was reported using modifications of the nuclear reactions for the ^{7}Li(d,n)2^{4}He reaction rates [49]. Apparent uncertainties still exist for this reaction used in BBN calculations. The latest data for the ^{7}Li(d,n)2^{4}He reaction rate focuses on three near-threshold ^{9}Be excited states. The contribution from a subthreshold resonance at 16.671 MeV in ^{9}Be was accounted for in Ref. [49]. The reaction rate for ^{7}Li(d,n)2^{4}He rate was found to become smaller than

Yields	Calculation	Observation
Y_p	0.2485+0.001-0.002	0.2453 ± 0.0034
D/H ($\times 10^{-5}$)	2.692+0.177-0.070	2.527 ± 0.03
^{3}He/H ($\times 10^{-6}$)	9.441+0.511-0.466	$\geq 11. \pm 2.$
^{7}Li/H ($\times 10^{-10}$)	4.283 +0.335-0.292	1.58+0.35 -0.28

Table 2: Standard BBN yields of light elements updated with input of recent experimental data for nuclear reactions. The table includes results based on astronomical observations. The mass fraction for ^{4}He is historically denoted by Y_p. It is taken from Ref. [39]. The deuterium abundance D/H = $(2.527 \pm 0.03) \times 10^{-5}$ [40, 41] is in accordance with $100\Omega_b h^2$ (BBN) = 2.225 ± 0.016 deduced from the observations of the cosmic microwave background [15]. The ^{3}He abundance is taken from Ref. [42], whereas the lithium abundance is from Ref. [23].

previous estimations by about a factor of 60 at typical temperatures considered during primordial nucleosynthesis. The new rates are shown to have a small impact on the final light element abundances in the standard BBN model. Only variations of the order of 0.002% compared with other calculations were found. But, for nonuniform density BBN models, the final ^{7}Li abundance increases up to 10% while those for other light nuclides with mass number $A > 7$ can increase by about 40% [49]. Thus, the cosmological lithium problem seems to be resilient from the standpoint of nuclear physics.

BBN predicts an isotopic ratio of $^6\text{Li}/^7\text{Li} \sim 10^{-5}$, whereas observation seemed to yield $^6\text{Li}/^7\text{Li} \sim 5 \times 10^{-2}$ [27]. Due to the apparent puzzle, a re-analysis of the reaction $^4\text{He}(d,\gamma)^6\text{Li}$ was made in Ref. [26], leading to new predictions for the angular distributions of the gamma-ray. A two-body potential model was adopted to calculate the cross section for this reaction at the BBN relevant energies [26, 50]. The potential parameters were fixed to reproduce experimental phase shifts and the measured ANCs reported in the literature. A good agreement with experimental data of the LUNA collaboration [51] was found. This work is again consistent with BBN predictions for the lithium isotopic ratio yielding $^6\text{Li}/^7\text{Li} = (1.5 \pm 0.3) \times 10^{-5}$. However, the non-detection of ^{6}Li in metal poor stars by the EXPRESSO collaboration [29] implies a simpler solution to the second cosmological lithium problem. According to Ref. [29] this problem never existed, because there is no evidence of ^{6}Li in Spite Plateau stars.

As for the puzzle concerning the ^{7}Li abundance, it does not seem possible that it could be solved by accurate laboratory experiments to measure nuclear reaction cross sections. Advanced modeling of nuclear astrophysical reactions also do not seem to solve the problem. But a considerable number of theoretical works to elucidate the lithium puzzle resort to ideas outside the range of nuclear physics. Perhaps, many physics processes could be different now than 13.8 billions years ago. Particles that could be decoupled from known ones, other sort of interactions, modifications in the values of fundamental constants, the possibility of non-standard BBN models, and other intriguing ideas have been explored and reported in the literature. Next we discuss a few ideas developed and tested by our group.

3 Dark parallel universes

One of the biggest puzzles in present day cosmology are the observations that most of the matter in the universe is composed of an obscure kind of matter, denoted as Dark Matter (DM). This DM seems to interact very weakly with the visible universe, mostly through gravitation. Astronomical observations of galaxy clusters dynamics and of anisotropies of the Cosmic Microwave Background (CMB) are the main proofs of the DM existence. Intensive experimental searches are being carried out to determine how DM might interact with visible matter other than by gravitation [52–54]. DM could be formed by some kind of Weakly Interacting Massive Particles (WIMPs), it could be composed of supersymmetric particles, or maybe some sort of sterile neutrinos, or even by other unknown particles.

Hypothetically, DM could lie in a separate mirror sector of particles composed of dark photons, dark electrons, etc. Such particles could interact in roughly the same way as Standard Model (SM) particles do, but only within their own sector. They could also interact weakly within and across sectors, the DM sector and visible matter sector [55–61]. Particle copies within the dark sectors do not necessarily have the equal masses and couplings as they do in the visible sector. All these thoughts open up a huge number of physics scenarios for DM and its interaction with what we see, the visible matter.

Present analysis of astronomical observations lead to a ratio of density parameters $\Omega_{DM}/\Omega_{visible} = 4.94 \pm 0.66$. This indicates that DM is about 5 times more massive than visible matter. This property has been used in Ref. [60–62] to study a possible scenario that our universe might consist of of 5 dark parallel universe sectors and not only one single ubiquitous sector. The motivation for this hypotheses is the assumption that all parallel universes contain nearly the same mass. A Weakly Interacting Massive Gauge Boson (WIMG) was proposed as a way to couple the different dark sectors and the ordinary matter with each dark universe. It was found that a massive WIMG, of the order of $E \sim 10$ TeV, does not change the properties of gravity and the SM. It was also shown that this WIMG model is consistent with BBN predictions and CMB observations, but it might have an impact on the lithium abundance.

The electroweak scale energy, i.e., the temperature of electroweak symmetry breaking, 159.5 ± 1.5 GeV, serves as an energy scale for comparisons. Much below this scale, particles in the model can be assumed to be massless. We group particles into matter/charge fields and a similar structure is used for DM. The generation of the WIMG mass is accomplished by means of a real scalar field, with a very short-range interaction supplied by the WIMG. Our formalism predicts that the number of dark sectors can have an overwhelming role which has been neglected in standard and non-standard BBN models. The obvious reason is that new degrees of freedom in the dark

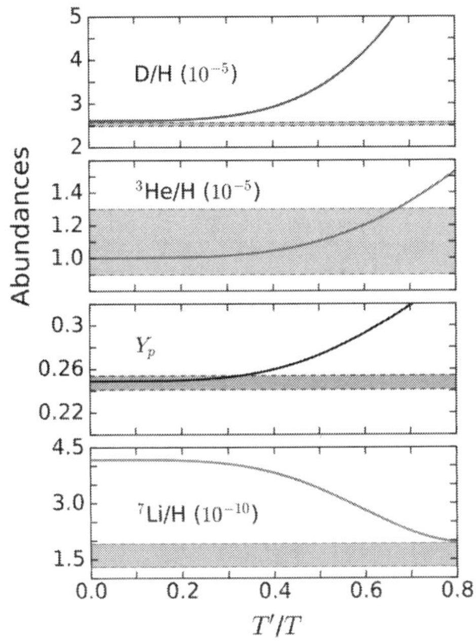

Figure 5: BBN model prediction for the relative abundances of D, ^{3}He, ^{4}He (mass fraction, Y_p) and ^{7}Li as a function of early universe temperatures ratio between the dark sector temperatures and visible matter temperature, T'/T. The calculations were done assuming that the number of dark sectors are $N_{DM} = 5$. The bands represent uncertainties in the observations [64].

sectors can considerably modify the early expansion rate of the universe [63] and consequently the light element abundances [60–62].

The radiation and entropy densities during the BBN epoch are related to the degrees of freedom of particles via $\rho(T) = (\pi^2/30)\, g_*(T)\, T^4$ and $s(T) = (2\pi^2/45)\, g_s(T)\, T^3$ with

$$g_*(T) = \sum_B g_B \left(\frac{T_B}{T}\right)^4 + \frac{7}{8}\sum_F g_F \left(\frac{T_F}{T}\right)^4,$$
$$g_s(T) = \sum_B g_B \left(\frac{T_B}{T}\right)^3 + \frac{7}{8}\sum_F g_F \left(\frac{T_F}{T}\right)^3, \quad (12)$$

with g_* and g_s being the number of degrees of freedom (d.o.f.), and $g_{B(F)}$ the fractions contributed by bosons (fermions) at the temperatures $T_{B(F)}$, while T is the temperature of the radiation thermal bath.

With minimal assumptions, only two temperatures, T for the ordinary matter sector, and T' for the dark sectors is used. It follows that the energy $\rho'(T')$ and the entropy $s'(T')$ densities within the dark sectors are also described via Eqs. (12) with the replacements $g_*(T) \to g'_*(T')$, $g_s(T) \to g'_s(T')$, and $T \to T'$. This implies that $x = (s'/s)^{1/3} \sim T'/T$ emerges from entropy conservation in all sectors. With each dark sector having the same total matter as the visible universe, it follows that $g_s(T_0) = g'_s(T'_0)$, leading to $x = T'/T$.

We recall that the Friedman equation is $H(t) = \dot{a}/a = \sqrt{(8\pi/3c^2)\, G_N\, \bar{\rho}}$, with $\bar{\rho}$ being the total energy density. Adding to it the matter from a number of dark sectors N_{DM}, the density becomes $\bar{\rho} = \rho + N_{DM}\, \rho'$. Thus, the Friedmann equation becomes $H(t) = 1.66\, \sqrt{\bar{g}_*(T)} T^2/M_{Pl}$, where $\bar{g}_*(T) = g_*(T)\left(1 + N_{DM}\, a\, x^4\right)$, $T(t)$ evolves with time, M_{Pl} denotes the Planck mass and $a = (g'_*/g_*)\,(g_s/g'_s)^{4/3} \sim 1$, for a not too small T'/T [63]. At about 1 MeV, standard BBN assumes $g_*(T = 1\text{ MeV}) = 10.75$.

With the additional dark particles, the number of d.o.f. becomes $\bar{g}_* = g_*\left(1 + N_{DM}\, x^4\right)$. The bounds of N_{DM} and x, or T'/T can be studied by comparing the results of BBN calculations including the new d.o.f. with the relative abundances of the light isotopes D, ^{3}He, ^{4}He, and ^{7}Li. Our result is displayed in Figure 5 with the abundances calculated as a function of T'/T for a fixed number of dark sectors $N_{DM} = 5$. The shaded bands displays the uncertainties related to the observed abundances. The observations of primordial elements of D, ^{3}He, and ^{4}He are compatible with early universe temperatures such that we get $T'/T \sim 0.2 - 0.3$.

The so-called ^{7}Li puzzle cannot be solved by invoking this model because if one uses $T'/T \sim 1$ it implies that the abundance of ^{7}Li becomes compatible with observations (see Fig. 5, bottom panel), but then the abundances of the other elements become in complete disagreement with observations. In fact, the d.o.f. factor $\bar{g}_*(T)$ depends much more strongly on T' than a variation of N_{DM}. For example, if one uses $T' = 0.3T$ for cold dark sectors, a large range of values for $N_{DM} = 1 - 50$ compatible with the D, ^{3}He and ^{4}He abundances is found [64].

Neutrinos play an important role for the counting of d.o.f. in the BBN. Figure 6 displays the BBN predictions for the primordial ^{4}He mass fraction as a function of extra number of neutrino families ΔN_ν in our parallel universe model, with $T' = 0.3T_{BBN}$ and $N_{DM} = 5$. The horizontal band represents the range of observed mass fraction [64]. The dark sectors model is therefore in accordance with a number of neutrino families $N_\nu = 3$. This result remains robust even assuming the existence of the parallel universes of dark matter and the changes in the number of d.o.f.

We conclude that the existence of multiple cold sectors is open to speculations, and a universe composed of visible matter and more than one sector of dark matter, e.g. $N_{DM} = 5$, and with temperatures of the dark sectors of the order of $T' \sim 0.2 - 0.3T$ does not influence the abundances of light elements. The burden of proof for the existence of a rather simple composition of the universe in terms of 5 dark parallel sectors plus a visible matter sector relies simply on us. Our inability to disprove it does not render our claim valid and it just adds to an incredible plethora of speculations about the physics of the early universe, as reported in the literature. As with the physics of neutron stars, BBN is a fertile ground for new theoretical

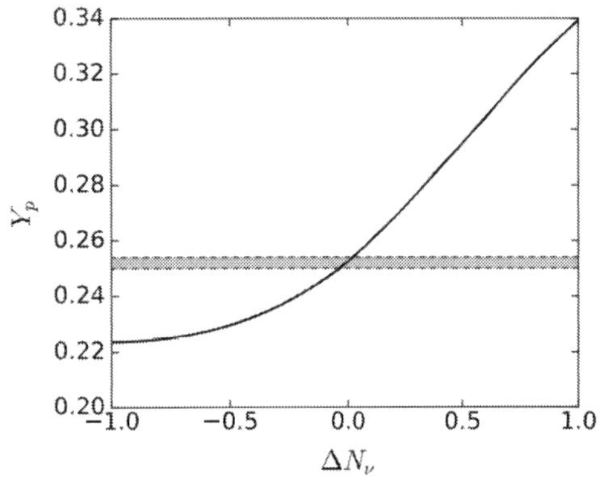

Figure 6: Predictions for the primordial ^{4}He mass fraction as a function of extra neutrino families, $N_\nu = 3 + \Delta N_\nu$, with $T' = 0.3T_{BBN}$ and $N_{DM} = 5$. The horizontal band represents the observed mass fraction [64].

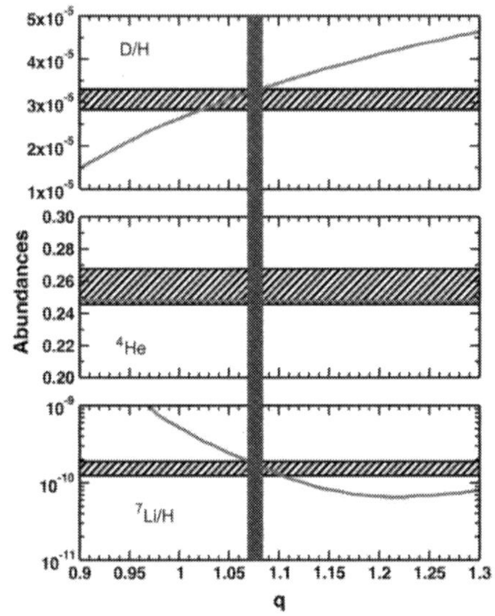

Figure 7: Predicted abundances D, ^{4}He and ^{7}Li (red curves) as a function of the Tsallis parameter q [75]. The observed primordial abundances including 1σ uncertainties are indicated by hatched horizontal bands [23, 39, 76]. The vertical (blue) band refers to the parameter q within the interval $1.069 < q < 1.082$.

ideas simply because only a few observations can be used as constraints.

4 The role of statistics for the BBN

The Maxwell-Boltzmann (MB) distribution is widely known to reproduce extremely well the distribution of velocities of particles in a thermal bath. The MB distribution is based on core assumptions adopted by the Boltzmann-Gibbs statistics. They are: (a) The time between collisions among particles is much longer than their interaction time; (b) the interactions are short-ranged; (c) no correlation exists between the particle velocities; (d) the collision energy is conserved without transfer to internal degrees of freedom. These are very constraining assumptions, not expected to be always valid even in a system of particles in thermodynamical equilibrium. In fact, alternatives to the Boltzmann-Gibbs (BG) statistics exist and are frequently used in the literature [65–67].

Our group has investigated the impact of using one of the so-called "non-extensive" statistics to describe the relative velocities of particles during the BBN [68]. Other ideas connected to the statistical aspects of BBN have been explored in the literature (see, e.g., Refs. [69, 70]). The non-extensive statistics implies that the entropies of two different microstates S_A and S_B of a system do not add in an extensive way (such as volume). That is, $S \neq S_A + S_B$. One of the most popular non-extensive statistics was reported in Refs. [66, 67]. We have calculated the effect on the BBN light element abundances using the so-called Tsallis statistics [68]. The Tsallis statistics depends on an additional parameter q representing its departure from the Boltzmann statistics. Nobody knows for sure what the physical meaning of this parameter is and maybe there is none, besides the fact that the Boltzmann statistics is recovered in the limit $q = 1$. But one knows that this statistics has been extremely successful in describing a large number of experimental data not in agreement with the Boltzmann statistics.

Previous applications of non-extensive statistics, seem to indicate that the non-extensive parameter q does not need to be very different than the Boltzmann value $q = 1$ to explain many physics phenomena not directly explainable with $q = 1$. Non-extensive Maxwellian velocity distributions have been used in the past to study stellar nuclear burning [71–74]. But in Ref. [68] it was first used for the reaction rate calculations during the BBN to obtain the ^{4}He, D, ^{3}He, and ^{7}Li abundances and study its impact on the lithium puzzle. The reaction rates for astrophysics are calculated using the average $\langle\sigma v\rangle$ around the Gamow window, where σ is the velocity dependent reaction cross section and v the ions relative velocit (see Eq.(7)). In Ref. [68] a non-extensive MB distribution for v was used in the BBN model. We have included reaction rates for p(n,γ)d, d(p,γ)^{3}He, d(d,n)^{3}He, d(d,p)t, ^{3}He(n,p)t, t(d,n)^{4}He, ^{3}He(d,p)^{4}He, ^{3}He(α,γ)^{7}Be, t(α,γ)^{7}Li, ^{7}Be(n,p)^{7}Li and ^{7}Li(p,α)^{4}He. These were fitted to cross sections based on presently available experimental data [68]. Our initial conclusion was that if either $q > 1$ or $q < 1$ the abundances of all elements were somehow affected, but the ^{7}Li abundance always increased [68]. Thus, we concluded that the lithium puzzle would always get worse if extensive statistics is used in the BBN model.

A small, but relevant physics input was neglected in Ref. [68]. This related to the proper modification of the reaction Q-values for backward reaction rates modified with the non-extensive statistics. Such a modification was considered in Ref. [75]. It led to amazing results with an impressive agreement with the observed abundances, including that of lithium, with a relatively small departure of the parameter q from unity. The H, D, ^{3}H, ^{3}He, and ^{4}He abundances remain basically unchanged, but the ^{7}Li abundance is appreciably modified, thus solving the ^{7}Li puzzle. The excellent agreement, shown in Fig. 7 for D, ^{4}He, and ^{7}Li, is obtained using $1.069 < q < 1.082$. One might argue that this agreement is fortuitous, if one is skeptical of non-extensive statistics. But this result shows that a possible solution to the cosmological lithium problem is due to a fine tuning of the physics involved in the BBN model.

In Fig. 7 the primordial abundances include 1σ uncertainties from observations, indicated with hatched horizontal bands [23, 39, 76]. The vertical (blue) band refers to the value of the parameter q for the already mentioned interval $1.069 < q < 1.082$. Ref. [75] was promoted as a research highlight of the American Astronomical Society [77], attesting the serious concern the Society has with the long standing lithium puzzle. It also displays a certain anxiety in the astronomy community that a reasonable physics explanation is long due, be it in the realm of stellar astration of light elements, cosmic ray spallation processes, or some other physics process. It is worthwhile mentioning that the agreement with D/H observations displayed in Fig. 7 would not be as good using the more recent compilation

$$(\mathrm{D/H})_{obs} = (2.55 \pm 0.03) \times 10^{-5}, \qquad (13)$$

based on an weighted average of 11 measurements [18].

The exercises played in Refs. [60–62, 68, 75] are based on ingenious ideas that some times are valid in order to explore a fine tuning of all possible physics processes. It is a trend that new physics ideas flourish when unexpected results emerge from experiments and observations.

Acknowledgements

This work was partially supported by the U.S. DOE grant DE- FG02-08ER41533.

References

[1] George Gamow, Phys. Rev. 70, 572 (1946).
[2] Ralph A. Alpher, Phys. Rev. 74, 1577 (1948).
[3] Ralph A. Alpher, Hans Bethe, and George Gamow, Phys. Rev. 73, 803 (1948).
[4] George Gamow, Nature 162, 680 (1948).
[5] Ralph A. Alpher, James W. Follin Jr, and Robert C. Herman, Phys. Rev. 92, 1347 (1953).
[6] Fred Hoyle and Roger J. Tayler, Nature 203, 1108(1964).
[7] P. James E. Peebles, Phys. Rev. Lett. 16, 410 (1966).
[8] G. Steigman, Ann. Rev. Nuc. Part. Phys. 57, 463 (2007).
[9] B.D. Fields, P. Molaro and S. Sarkar, (Particle Data Group), Prog. Theor. Exp. Phys. 2022, 083C01 (2022).
[10] C.A. Bertulani and T. Kajino, Prog. Part. Nucl. Phys. 89, 56 (2016).
[11] Richard H. Cyburt, Brian D. Fields, Keith A. Olive, and Tsung-Han Yeh, Rev. Mod. Phys. 88, 015004 (2016).
[12] Cyril Pitrou, Alain Coc, Jean-Philippe Uzan, Elisabeth Vangioni, Phys. Reports 754, 1 (2018).
[13] Brian D. Fields, Keith A. Olive, Tsung-Han Yeh, Charles Young, JCAP 03, 10 (2020).
[14] Michael S. Turner, arXiv:2111.14254 (2022).
[15] N. Aghanim et al, A&A 641, A6 (2020).
[16] The ALEPH Collaboration, The DELPHI Collaboration, The L3 Collaboration, The OPAL Collaboration, The SLD Collaboration, The LEP Electroweak Working Group, The SLD Electroweak and Heavy Flavor Groups, Phys. Reports 427, 257 (2006).
[17] F.M. Gonzalez et al., Phys. Rev. Lett. 127, 162501 (2021).
[18] Tsung-Han Yeh, Keith A. Olive and Brian D. Fields, JCAP 03, 046 (2021).
[19] Alain Coc, Jean-Philippe Uzan and Elisabeth Vangioni, JCAP 10, 50 (2014) .
[20] M. Spite and F. Spite, Nature 297, 483 (1982); F. Spite and M. Spite, A&A 115, 357 (1982).
[21] Brian D. Fields and Keith A. Olive, arXiv:2204.03167.
[22] W. Aoki, Paul S. Barklem, Timothy C. Beers, Norbert Christlieb, Susumu Inoue, Ana E. García Perez, John E. Norris, Daniella Carollo, Ap. J. 698, 1803 (2009).
[23] L. Sbordone, et al., A&A 522, A26 (2010).
[24] A. Coc, Stephane Goriely, Yi Xu, Matthias Saimpert, Elisabeth Vangioni, Ap. J. 744, 158 (2012).
[25] R.G. Pizzone R. Sparta, C. A. Bertulani, C. Spitaleri, M. La Cognata, J. Lalmansingh, L. Lamia, A. Mukhamedzhanov, A. Tumino, Ap. J. 786, 112 (2014).
[26] A. M. Mukhamedzhanov, Shubhchintak, C.A. Bertulani, Phys. Rev. C 93, 045805 (2016).
[27] M. Asplund, D. L. Lambert, P. E. Nissen, F. Primas and V. V. Smith, Ap. J. 644, 229 (2006).
[28] K. Lind, J. Melendez, M. Asplund, R. Collet and Z. Magic, A&A 554, A96 (2013).
[29] Ella Xi Wang, Ella Xi et al., Monthly Notices of the Royal Astronomical Society 509, 1521 (2021).
[30] R.V. Wagoner, Ap. J. Suppl. Ser. 18, 247 (1969).
[31] L. Kawano, "Let's Go: Early Universe. Primordial Nucleosynthesis: The Computer Way"; NASA Technical Reports Server (NTRS): Hampton, VA, USA, 1992.
[32] C.A. Bertulani, F. W. Hall, B. I. Santoyo, unpublished (2022).
[33] E.E. Salpeter, Aust. J. Phys. 7, 373 (1954); E.E. Salpeter and H.M. Van Horn, Ap. J. 155, 183 (1969).
[34] B. Wang, C.A. Bertulani, and A.B. Balantekin, Phys. Rev. C 83, 018801 (2011).
[35] M.A. Famiano, A.B. Balantekin and T. Kajino Phys. Rev. C 93, 045804 (2016).
[36] Yudong Luo, Michael A. Famiano, Toshitaka Kajino, Motohiko Kusakabe, and A. Baha Balantekin, Phys. Rev. D 101, 083010 (2020)
[37] C. Spitaleri, C.A. Bertulani, L. Fortunato, A. Vitturi, Phys. Lett. B 755, 275 (2016).

[38] C.A. Bertulani and C. Spitaleri, Eur. Phys. J., 165, 02002 (2017).

[39] Erik Aver, Danielle A. Berg, Keith A. Olive, Richard W. Pogge, John J. Salzer and Evan D. Skillman, JCAP 03, 027 (2021).

[40] Signe Riemer-Sorensen and Espen Sem Jenssen, Universe 2, 44 (2017). doi:10.3390/universe3020044.

[41] R.J. Cooke, M. Pettini, and C.C. Steidel, ApJ 855, 102 (2018).

[42] T. M. Bania, R. T. Rood, D. S. Balser, Nature 415, 54 (2002).

[43] C. Broggini, L. Canton, G. Fiorentini, F. L. Villante, JCAP 6, 30 (2012).

[44] S. Q. Hou, J. J. He, S. Kubono, and Y. S. Chen, Phys. Rev. C 91, 055802 (2015).

[45] M. Barbagallo, et al., Phys. Rev. Lett. 117, 152701 (2016).

[46] T. Kawabata et al., Phys. Rev. Lett. 118, 052701 (2017).

[47] L. Lamia, C. Spitaleri, C. A. Bertulani, S. Q. Hou, M. La Cognata, R. G. Pizzone, S. Romano, M. L. Sergi, and A. Tumino, Ap. J. 850, 175 (2017).

[48] M. Hartos, C. A. Bertulani, Shubhchintak, A. M. Mukhamedzhanov, and S. Hou, Astrophys. J. 862, 62 (2018).

[49] S. Q. Hou, T. Kajino, T. C. L. Trueman, M. Pignatari, Y. D. Luo, C. A. Bertulani, Astrophysical J. 920, 145 (2021).

[50] C.A. Bertulani, Comput. Phys. Commun. 156, 123 (2003).

[51] M. Anders et al., Phys. Rev. Lett. 113, 042501 (2014).

[52] J. L. Feng, Ann. Rev. Astr. Astrophys., 48 495 (2010).

[53] G. Bertone, D. Hopper, and J. Silk, Phys. Rep. 405, 279 (2005).

[54] G. Bertone, Nature 468, 389 (2010).

[55] T.D. Lee and C. N. Yang, Phys. Rev. 104, 254 (1956).

[56] Y. Kobzarev, L. Okun, I. Pomeranchuk, Yad. Fiz. 3, 1154 (1966).

[57] M. Pavsic, Int. J. Theor. Phys. 9, 229 (1974).

[58] R. Foot, H. Lew, R. Volkas, Phys. Lett. B 272, 67 (1991).

[59] E. Akhmedov, Z. Berezhiani, G. Senjanovic, Phys. Rev. Lett. 69, 3013 (1992).

[60] O. Oliveira, C. A. Bertulani, M. S. Hussein, W. de Paula, T. Frederico, arXiv:1108.2723 (2011).

[61] O. Oliveira, C. A. Bertulani, M. S. Hussein, W. de Paula, T. Frederico, Braz. J. Phys. 46, 721 (2016).

[62] C.A. Bertulani, T. Frederico, J. Fuqua, M.S. Hussein, O. Oliveira, W. de Paula, AIP Proc.1498, 134 (2013).

[63] Z.G. Berezhiani, A.D. Dolgov, R.N. Mohapatra, Phys. Lett. B375, 26 (1996).

[64] C.A. Bertulani, V. Challa, J.J. He, S.Q. Hou, Ravinder Kumar, arXiv:1704.07803.

[65] A. Rényi, Proc. of the 4th Berkeley Symposium on Mathematics, Statistics and Probability, p. 547, 561 (1960).

[66] C. Tsallis, J. Stat. Phys. 52, 479 (1988).

[67] M. Gell-Mann and C. Tsallis, eds., "Nonextensive Entropy-Interdisciplinary Applications, (Oxford University Press, New York, 2004). For a regularly updated bibliography see URL: http://tsallis.cat.cbpf.br/biblio.htm

[68] C.A. Bertulani, J. Fuqua and M.S. Hussein, Ap. J. 767, 67 (2013).

[69] M.Yu. Khlopov, Yu.L. Levitan, E.V. Sedelnikov and I.M. Sobol, Yadernaya Fizika 57, 1466 (1994). [English translation: Phys. Atom. Nucl. 57, 1393 (1994)]

[70] E.V. Sedelnikov, S.S. Filippov and M.Yu. Khlopov, Yadernaya Fizika 58, 280 (1995). [English translation: Phys. Atom. Nucl. 58, 235 (1995)]

[71] M. Lissia and P. Quarati, Europhys. News 36, 211 (2005).

[72] J. Haubold and D. Kumar, Astroparticle Phys. 29 ,70 (2008).

[73] S. Degl'Innocenti et al, Phys. Lett. B 441, 291 (1998).

[74] M. Coraddu et al., Braz. J. Phys. 29, 153 (1999).

[75] S.Q. Hou, J.J. He, A. Parikh, D. Kahl, C.A. Bertulani, T. Kajino, G.J. Mathews and G. Zhao, Ap. J. 834, 165 (2017).

[76] K.A. Olive, P. Petitjean, E. Vangioni, J. Silk, MNRAS 426, 1427 (2012).

[77] Susanna Kohler, AAS-NOVA, "Fixing the Big Bang Theory's Lithium Problem", 15 February (2017).

Recoil mass separators for nuclear astrophysics: the role of ERNA

Raffaele Buompane[1,2,*] *Giuseppe* Porzio[1,2,] *Claudio* Santonastaso[1,2,]

[1]Dipartimento di Matematica e Fisica, Università degli Studi della Campania "L. Vanvitelli", Viale Lincoln, 5, 81100 Caserta, Italy;
[2]Istituto Nazionale di Fisica Nucleare, Sezione di Napoli, Via Cintia snc, 80126 Napoli, Italy

Abstract. The measurements of radiative capture reactions can be performed in inverse kinematic detecting directly the recoil produced in the nuclear reaction using a recoil mass separator (RMS). The development of RMS allows the possibility to overtake both the problems of gamma background signal and purity and production of target. The European Recoil Separator for Nuclear Astrophysics (ERNA) is a RMS designed with the main goal of determining the $^{12}C(\alpha,\gamma)^{16}O$, presently hosted at Center for Isotopics Research and Cultural hEritage laboratory of Department of Mathematics and Physics, University of Campania Luigi Vanvitelli, Caserta Italy (CIRCE-DMF). A general discussion on measurement techniques with recoil mass separator will be presented in this contribution with a focus on the ERNA one.

1 Introduction

The radiative capture reactions (α,γ) and (p,γ) are involved in stellar evolution and nucleosynthesis in several astrophysical scenarios [1, 2]. Their paramount importance in nuclear astrophysics triggered several experiments devoted to measurements where the direct experimental information on radiative capture reactions has been collected measuring the reaction yields through the observation of the emitted γ-rays [3, 4]. Due to the lower cross section at astrophysical energies of interest, usually, the gamma counting rate is too low with respect to the environmental and instrumental detector background. The signal to background ratio limitation stimulated the developments of two different approaches to perform direct measurements: one consists in performing experiments underground, the other using RMS. Thanks to the underground installation, the environmental background is reduced by several orders of magnitude[5]. The resulting enhancement of the signal to noise ratio allows lower energies measurements where fewer reaction events are expected. Depending on the particular reaction, background contributions may still be induced by the beam and impurity in the target. The RMS allows the radiative capture reaction measurements with the direct detection of the recoil, overtaking the background problem. The RMS measurements are performed in inverse kinematic with the advantage to have completely different systematic uncertainty with respect to the direct measurements.

*e-mail: raffaele.buompane@unicampania.it

The first RMS devoted to the measurement of nuclear cross section of astrophysical interest was built at Caltech [6], although the limited beam suppression imposed to detect recoils in a coincidence with capture γ-rays in NaI detectors at the target position. Later, the NABONA Collaboration (NAples BOchum Nuclear Astrophysics) built and commissioned a RMS devoted to the measurement of the ^{7}Be(p, γ)^{8}B cross-section relying only on recoils detection, using a windowless hydrogen gas target and a ^{7}Be beam [7]. Building upon this successful experience, the same groups started a new project to realize the ERNA RMS at the 4MV Dynamitron Tandem Laboratorium of the Ruhr-Universität Bochum (Germany) [8, 9, and reference therein], with the aim of studying the ^{12}C(α,γ)^{16}O reaction. In the following years, up to nowadays, several RMS devoted to the nuclear astrophysics study are commissioned around the world, more information about the other RMS are reported in [10, and reference therein].

2 Principles of RMS experiments

In a radiative capture reaction, the fusion of projectile and target particles produces a nucleus that recoils in the laboratory frame system with a momentum quite similar to the one of the projectile [11]. The energy spectrum of the recoils is included in the range $[E_{r,1} - \Delta E_\gamma, E_{r,0} + \Delta E_\gamma]$, where $E_{r,0}$ and $E_{r,1}$ are the energy of the recoil produced at beginning and at the end of the target, respectively, while ΔE_γ is the maximum energy shift of the recoils due to the emission of a γ-ray with energy $E_\gamma = Q + E_{cm}$, given by

$$\Delta E_\gamma = \frac{E_\gamma^2}{2 \cdot M_r \cdot c^2} + \frac{E_\gamma}{c^2} \cdot \sqrt{\frac{2 \cdot E_b}{M_r}}, \tag{1}$$

The momentum transfer in γ-ray emission produces a recoil angle up to a maximum angle $\theta_{\gamma,max}$ given by

$$\theta_{\gamma,max} = arctan\frac{E_\gamma}{c \cdot p_{cm,lab}}, \tag{2}$$

where $p_{cm,lab}$ is the momentum of the center-of-mass in the laboratory reference system.

This means that the secondary beam composed by the recoils will be diverging right after the target and needs to be focused toward the beam axis to be efficiently transported further. Measuring in inverse kinematics decreases the divergence of the beam in the laboratory frame facilitating the recoils transmission. Typical recoils to projectile ratios ranges from 10^{-10} to 10^{-17} or lower, depending on the cross-section under study. In order to count the recoils in a detector, most of the primary beam (projectiles particles) need to be filtered out after the target. Therefore, an RMS consists of a series of filters and focusing elements that allows for an efficient suppression of the primary beam and ensures the transport of all recoils (in a given charge state) at the end detector such that they can be identified and counted.

The number of recoils N_r observed in the final detector, having an efficiency ϵ, is given by [1]

$$N_r = N_b \cdot T_{RMS} \cdot \Phi_{q_r} \cdot \epsilon \cdot \int_{E_b - \Delta E_b}^{E_b} \frac{\sigma(E)}{|\mathrm{d}E/\mathrm{d}N_t|} \mathrm{d}E \ , \tag{3}$$

where N_b and $T_{RMS}(q)$ are, respectively, the number of projectiles impinging on the target and the transmission of the recoils from the target to the final detector, when the recoil charge state q_r with probability Φ_{q_r} is selected.

The interaction cross-section $\sigma(E)$ is integrated over the beam energy loss in the target ΔE_b at the beam energy E_b that is determined by the stopping power of the beam ions in

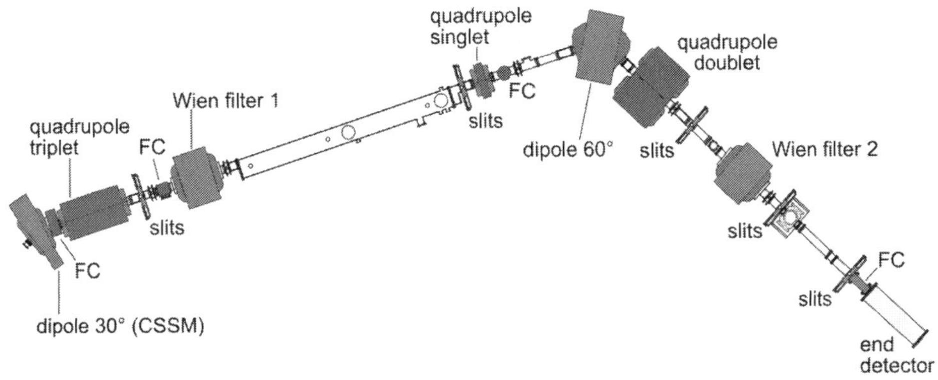

Figure 1. Layout of the ERNA RMS at the CIRCE-DMF Tandem Accelerator Laboratory. CSSM indicates the charge state selection magnet; FC indicates the Faraday cups along the separator [10].

the target dE/dN_t, where N_t is the number of nuclei per unit target area. Of course, an accurate and precise determination of $\sigma(E)$ relies on an adequate estimate of all quantities in Equation (3).

The full acceptance of recoils is a crucial issue in experiments with mass separators. Corrections for missing acceptance of recoils depend critically upon their actual distribution in the phase space [12], that is difficult to predict depending on the details of target effects and γ-ray angular distribution. The most effective method to know in detail the acceptance of the separator, i.e., the capability of the RMS to transport to the end detector the recoils regardless of their angle and energy, is to directly measure it. This can be done using a pilot beam mimicking the recoils beam properties, achieved using an electrostatic deflection unit, and varying the beam energy to sample the emittance of the recoils and determine their transmission through the RMS, as reported in [9, 13].

3 The ERNA separator at CIRCE

In 2009, ERNA was moved to the Tandem Accelerator Laboratory of the CIRCE-DMF, University of Campania, Caserta, Italy, where a 3.0 MV NEC 9SDH-2 Pelletron accelerator is located [14]. This has provided the opportunity for an upgrade in the separator's layout. In particular, a further charge filtering element, the Charge State Selection Magnet (CSSM in the following), has been placed after the gas target and before the first focusing element of the separator. The layout of the separator is shown in Figure 1.

3.1 Experimental setup

Two 40-sample MC-SNICS (Multicathode source of Negative Ions by Cesium Sputtering) Cs-sputter negative ion sources are available. One, with the spherical ionizer option and inlet gas, is used for the production of stable species ion beams and AMS (accelerator mass spectrometry) measurements. A second one is dedicated to the production of offline radioactive ion (RI) beams, in particular ^{7}Be [15]. This offers the opportunity to investigate processes with both stable and RI beams.

Depending on the reaction under study, the separator has been equipped with different windowless gas targets: extended ^{4}He and ^{3}He target for the study of the $^{12}C(\alpha,\gamma)^{16}O$ and $^3He(\alpha,\gamma)^7Be$, respectively [8, 13], another different extended ^{4}He one for the study of ^{15}N$(\alpha, \gamma)^{19}$F [16], an extended H_2 target for the measurement of the ^{7}Be(p, $\gamma)^{8}$B cross-section [17], and a supersonic jet ^{4}He one [18] in view of the measurement of the ^{12}C$(\alpha, \gamma)^{16}$O at even lower energies. The gas targets are provided with an additional Ar gas layer (poststripper) so that recoils reach charge state equilibrium regardless of the reaction position within the target.

As said, an RMS is an alternating series of filters and focusing elements. The first filtering element of ERNA is the CSSM placed at the shortest distance allowed by geometrical constraints from the target. Its role is to select a single charge state for both projectile and recoil ions before the actual mass separation takes place. The next filtering element is the first velocity filter (Wien filter 1) that removes most of the primary beam thanks to the fact that projectiles and recoils have similar momenta and thus different velocities. The other filters are the 60° dipole magnet and the second velocity filter (Wien filter 2) that suppress the part of the primary beam that for different reasons may pass Wien filter 1, e.g., scattering in the target area. The recoils beam focusing is provided by the magnetic quadrupole triplet, the quadrupole singlet, and the quadrupole doublet. Several FCs, slit systems, wire scan, apertures, and a magnetic steerer (not shown in Figure 1) are installed along the RMS beam line for setting-up and diagnosis purposes. Depending on the capture reaction characteristics, the end detector can be either a ΔE-E ionization chamber telescope (ICT) or a Time Of Flight Energy (TOF-E) detection system. For the measurements of ^{12}C$(\alpha, \gamma)^{16}$O angular distribution a gamma array based on NaI detectors has been developed and installed at target position.

ERNA collaboration is also involved in the study of nuclear reactions in direct kinematics which does not involve the RMS system. For this purpose an array of GASTLY (GAs-Silicon Two-Layer sYstem) detectors [19] for the measurements of reactions with charged particles in the exit channel. The use of the GASTLY array allowed for unambiguous identification of both protons and alpha particles.

ERNA is fully computer-controlled, which allows, through sophisticated algorithms, to tune the beam both in automatic and manual mode always supervisionated by interlocks that define the right restrictions and, in the end, ensures long-term stability.

3.2 Scientific results

The ERNA collaboration performed successfully several measurements and proved to be a valuable tool in the measurement of the total cross-section of radiative capture reactions of astrophysical interest. The first results were from the measurement of the total cross-section of the ^{12}C$(\alpha, \gamma)^{16}$O reaction [20]. This measurement significantly extended the energy range of the measurements available at that time with a considerable improvement on the overall precision. It was possible to get a new insight on the relevance of the cascade transitions that were later investigated by DRAGON [21] and ERNA [22].

The use of ERNA RMS for the measurement of the ^{3}He$(\alpha, \gamma)^{7}$Be reaction cross-section also led to relevant results. Besides giving an important contribution to solve the then long-standing issue of the difference between extrapolation based on experiments selected according to the measurement method, the results indicated that at energies above $E_{\mathrm{c.m.}} \simeq 1$ MeV the energy dependence was significantly different from the one foreseen by the most adopted models [13].

The ERNA RMS was used to measure the reaction yield of the ^{15}N$(\alpha, \gamma)^{19}$F resonances at $E_{\mathrm{c.m.}} = 1323$ and 1487 keV. For this purpose, enhanced intensity of N beam from a SNICS source was thoroughly investigated [23]. The yield of the two resonances was observed as a function of the energy, thus, from the convolution of the cross-section expressed as a Breit–Wigner function and the target density profile, the widths Γ_γ and Γ_α could be extracted for both resonances. A detailed description of the analysis is given in [16].

More recently, the ERNA Collaboration completed the measurement of the ^{7}Be(p,γ)^{8}B total cross-section in the energy range $E_{\rm eff}$=367.2 keV to 812.2 keV using the radioactive ^{7}Be beam available at CIRCE-DMF with intensities up to 10^9 pps [15]. The measurement performed using the inverse kinematics approach, provided for the first time a results with adequate precision to determine the cross-section of ^{7}Be(p,γ)^{8}B at astrophysical energy. The experimental apparatus has been described in detail in [17, 24]. A detailed description of the data analysis is presented in [25] and its supplemental material.

As said, ERNA collaboration is involved not only in the RMS measurements. In the last years, studies of other nuclear reactions in direct kinematics, apparatuses developments and measurements of astrophysical interest have been performed [19, 26–30]. Recently, the ^{12}C+^{12}C reaction was studied using a partial configuration available in the first development phase of the GASTLY detector array, i.e., four GASTLY detectors with the SSD as an entire pad. The thick target yield has been measured in the energy range $E_{\rm c.m.}$=2.51 – 4.36 MeV and the astrophysical S-factors of ^{12}C(^{12}C,p)^{23}Na and ^{12}C(^{12}C,α)^{20}Ne have been extracted [28].

4 Conclusion

RMS allow to detect directly the recoils produced in a fusion process, overtaking the problem of the unfavorable signal to background ratio, typical in nuclear astrophysics experiments. The RMS measurements are usually performed in inverse kinematic, hence, the target, beam and detection system are completely different with respect to standard technique in direct kinematic, this assures the possibility to explore very different systematic errors. ERNA is the unique European RMS devoted to nuclear astrophysics measurements and is actually hosted at CIRCE-DMF laboratory, Caserta. The ERNA separator proved to be a valuable tool in the measurement of the total cross-section of radiative capture reactions of astrophysical interest. The experimental program of the ERNA collaboration aims at the extension at low energy ranges of the ^{12}C(α,γ)^{16}O and ^{15}N(α,γ)^{19}F cross-sections, and of the ^{7}Be(p,γ)^{8}B at E_{cm} > 1 MeV. Similarly, also the charged particle spectroscopy measurements will continue with the investigation of the ^{12}C+^{12}C and ^{12}C+^{16}O fusion process extending the energy range measured in the last few years. Owing to the good performances of the ERNA RMS and GASTLY, several other reactions can be envisioned in the longer term.

References

[1] C.E. Rolfs, W.S. Rodney, *Cauldrons in the Cosmos: Nuclear Astrophysics* (The University of Chicago Press, 1988)
[2] C. Iliadis, *Nuclear Physics of Stars* (Wiley-VCH Verlag GmbH & Co. KGaA, Weinheim, Berlin, Germany, 2007)
[3] C. Rolfs, C. Barnes, Annu. Rev. Nucl. Part. Sci. **40**, 45 (1990)
[4] C.R. Brune, B. Davids, Annu. Rev. Nucl. Part. Sci. **65**, 87 (2015)
[5] D. Bemmerer, F. Confortola, A. Lemut, R. Bonetti, C. Broggini, P. Corvisiero, H. Costantini, J. Cruz, A. Formicola, Z. Fülöp et al., The European Physical Journal A - Hadrons and Nuclei **24**, 313 (2005)
[6] M. Smith, C. Rolfs, C. Barnes, Nuclear Instruments and Methods in Physics Research Section A: Accelerators, Spectrometers, Detectors and Associated Equipment **306**, 233 (1991)
[7] L. Gialanella, F. Strieder, K. Brand, L. Campajola, A. D'Onofrio, U. Greife, E. Huttel, F. Petrazzuolo, V. Roca, C. Rolfs et al., Nucl. Instr. Meth. A **376**, 174 (1996)

[8] L. Gialanella, D. Schürmann, F. Strieder, A. Di Leva, N. De Cesare, A. D'Onofrio, G. Imbriani, J. Klug, C. Lubritto, A. Ordine et al., Nucl. Instr. Meth. A **522**, 432 (2004)
[9] D. Schürmann, F. Strieder, A. Di Leva, L. Gialanella, N. De Cesare, A. D'Onofrio, G. Imbriani, J. Klug, C. Lubritto, A. Ordine et al., Nucl. Instr. Meth. A **531**, 428 (2004)
[10] R. Buompane, A. Di Leva, L. Gialanella, G. Imbriani, L. Morales-Gallegos, M. Romoli, Universe **8** (2022)
[11] C. Ruiz, U. Greife, U. Hager, Eur. Phys. J. A **50**, 99 (2014)
[12] L. Gialanella, D. Schürmann, PoS **ENAS 6**, 058 (2013)
[13] A. Di Leva, M. De Cesare, D. Schürmann, N. De Cesare, A. D'Onofrio, L. Gialanella, R. Kunz, G. Imbriani, A. Ordine, V. Roca et al., Nucl. Instr. Meth. A **595**, 381 (2008)
[14] F. Terrasi, D. Rogalla, N. De Cesare, A. D'Onofrio, C. Lubritto, F. Marzaioli, I. Passariello, M. Rubino, C. Sabbarese, G. Casa et al., Nucl. Instr. Meth. B **259**, 14 (2007)
[15] B.N. Limata, L. Gialanella, A.D. Leva, N.D. Cesare, A. D'Onofrio, G. Gyurky, C. Rolfs, M. Romano, D. Rogalla, C. Rossi et al., Nucl. Instr. Meth. B **266**, 2117 (2008)
[16] A. Di Leva, G. Imbriani, R. Buompane, L. Gialanella, A. Best, S. Cristallo, M. De Cesare, A. D'Onofrio, J.G. Duarte, L.R. Gasques et al., Phys. Rev. C **95**, 045803 (2017)
[17] D. Schürmann, A. Di Leva, L. Gialanella, M. De Cesare, N. De Cesare, G. Imbriani, A. D'Onofrio, M. Romano, M. Romoli, F. Terrasi, Eur. Phys. J. A **49**, 80 (2013)
[18] D. Rapagnani, R. Buompane, A. Di Leva, L. Gialanella, M. Busso, M. De Cesare, G. De Stefano, J. Duarte, L. Gasques, L. Morales-Gallegos et al., Nucl. Instr. Meth. B **407**, 217 (2017)
[19] M. Romoli, L. Morales-Gallegos, M. Aliotta, C.G. Bruno, R. Buompane, A. D'Onofrio, T. Davinson, M. De Cesare, A. Di Leva, P. Di Meo et al., Eur. Phys. J. A **54**, 142 (2018)
[20] D. Schürmann, A. Di Leva, L. Gialanella, D. Rogalla, F. Strieder, N. De Cesare, A. D'Onofrio, G. Imbriani, R. Kunz, C. Lubritto et al., Eur. Phys. J. A **26**, 301 (2005)
[21] C. Matei, L. Buchmann, W.R. Hannes, D.A. Hutcheon, C. Ruiz, C.R. Brune, J. Caggiano, A.A. Chen, J. D'Auria, A. Laird et al., Phis. Rev. Lett. **97**, 242503 (2006)
[22] D. Schürmann, A. Di Leva, L. Gialanella, R. Kunz, F. Strieder, N. De Cesare, M. De Cesare, A. D'Onofrio, K. Fortak, G. Imbriani et al., Phys. Lett. B **703**, 557 (2011)
[23] A. Di Leva, A. Pezzella, N. De Cesare, A. D'Onofrio, L. Gialanella, M. Romano, M. Romoli, D. Schürmann, F. Terrasi, G. Imbriani, Nucl. Instr. Meth. A **689**, 98 (2012)
[24] R. Buompane, N. De Cesare, A. Di Leva, A. D'Onofrio, L. Gialanella, M. Romano, M. De Cesare, J.G. Duarte, Z. Fülöp, L. Morales-Gallegos et al., Eur. Phys. J. A **54**, 92 (2018)
[25] R.Buompane, A. Di Leva, L. Gialanella, A. D'Onofrio, M. De Cesare, J.G. Duarte, Z. Fülöp, L.R. Gasques, G. Gyürky, L. Morales-Gallegos et al., Phys. Lett. B p. 136819 (2021)
[26] J. Zickefoose, A.D. Leva, F. Strieder, L. Gialanella, G. Imbriani, N.D. Cesare, C. Rolfs, J. Schweitzer, T. Spillane, O. Straniero et al., Phys. Rev. C **97** (2018)
[27] L. Morales-Gallegos, M. Aliotta, C. Bruno, R. Buompane, T. Davinson, M. De Cesare, A. Di Leva, A. D'Onofrio, J. Duarte, L. Gasques et al., Eur. Phys. J. A **54** (2018)
[28] L. Morales-Gallegos, M. Aliotta, L. Gialanella, A. Best, C. Bruno, R. Buompane, T. Davinson, M. De Cesare, A. Di Leva, A. D'Onofrio et al., Eur. Phys. J. A **58** (2022)
[29] F. Terrasi, F. Marzaioli, R. Buompane, I. Passariello, G. Porzio, M. Capano, S. Helama, M. Oinonen, P. Nöjd, J. Uusitalo et al., Radiocarbon **62**, 1403–1418 (2020)
[30] C. Santonastaso, R. Buompane, A. Di Leva, L. Morales-Gallegos, N. Itaco, G. Landi, H. Neitzert, D. Rapagnani, L. Gialanella, Nuovo cimento **100**, 44 (2021)

Magnetic and non-magnetic AGB mixing for *s*-processing

Maurizio Busso[1,2,*]

[1]Department of Physics and Geology, University of Perugia, Via A. Pascoli snc, I-06125 Perugia, Italy
[2]INFN Section of Perugia, Via A. Pascoli snc, I-06125 Perugia, Italy

Abstract. I outline a few features of recent models for the formation of the neutron source $^{13}C(\alpha,n)^{16}O$ in low mass stars ($1 \lesssim M/M_{\odot} \lesssim 3$, *LMS*) ascending for the second time the Red Giant Branch, generally called *Asymptotic Giant Branch*, or *AGB* stars. I also briefly outline the nucleosynthesis results obtained trough them. The mentioned models consider the physical structure below the frequent downward extensions of the convective envelope into the He-intershell (the so-called *third dredge-up* or *TDU* episodes). There, the conditions are such that the occurrence of further mixing is strongly facilitated, due to the minimal temperature gradient. A way to induce proton mixing from the envelope (certainly not the only one) arises whenever the ambient magnetic fields expected for LMS promote the buoyancy of strongly magnetized flux tubes. I review some characteristics of the ensuing mixing episodes, mentioning how different hydrodynamical processes might yield similar effects, thus encouraging stellar physicists to verify in more detail this possibility.

1 Introduction

In another lesson at this school [1] the general picture of neutron-capture nucleosynthesis was layed down and due space was dedicated to the slow (*s*) process of neutron addition, occurring in low mass stars ($1 \lesssim M/M_{\odot} \lesssim 3$). There, the involved evolutionary stages are those of the so-called *Asymptotic Giant Branch*, when the repeated activation of the neutron sources $^{13}C(\alpha,n)^{16}O$ and $^{22}Ne(\alpha,n)^{25}Mg$ make moderate neutron fluxes available, feeding nucleosynthesis along the valley of β stability [2, 3].

Of the two mentioned neutron sources, the first is the dominating one, but its activation is not straightforward in current stellar models, as it requires that considerable amounts of ^{13}C be available in the He-rich layers of the star, while previous shell-H burning through the CNO cycle is expected to consume carbon to minimal concentrations.

Actually, ^{12}C is abundantly reproduced in the relevant layers, thanks to He-burning occurring in a thin shell above the border of a degenerate C-O core. Then, sudden variations in the thermal properties (*Thermal Pulses*), due to instabilities occurring in *thin shell* nuclear burning [4, 5] spread the produced carbon throughout the He-rich zone (at an abundance of about 20 – 25%). Hence, what is really needed is just a mechanism introducing protons from the envelope into those layers, during the *TDU* episodes [1]. They would then be rapidly captured by ^{12}C via:

$$^{12}C(p,\gamma)^{13}N(\beta^{+}\nu)^{13}C \quad (1)$$

*e-mail: maurizio.busso@pg.infn.it

For several years the required mixing processes remained rather mysterious and their efficiency was simply parameterized [2, 6]. They need to involve a concentration of protons small enough to avoid that, after reaction (1), ^{13}C itself be efficiently destroyed by further proton captures, producing ^{14}N. Indeed, nuclear descendants of this isotope act as *neutron poisons*, preferentially absorbing rather than producing neutrons, thus making the whole mechanism inefficient.

In section 2 I report on an attempt at founding the ^{13}C formation on physical grounds, based on the known property of LMS to host magnetic fields, organized in flux tubes, maintaining a *stellar dynamo* process [7]. There, I also show some nucleosynthesis results obtained in that scenario, where a remarkable agreement with observations emerges. Basic properties of such a mixing model, suitable to be shared by other physical processes, are then presented in section 3, as a stimulous to other groups to propose complementary ways of driving slow neutron captures in AGB stars.

2 ^{13}C from magnetic buoyancy and the ensuing nucleosynthesis

Broadly speaking, it is self-evident that magnetic fields of considerable geometrical and evolutionary complexity are generated in complex electrically-conducting astrophysical fluids [8]. The explanation of the huge observational evidence that magnetic fields persist in LMS for extremely long times requires that, in the induction equation:

$$\frac{\partial \vec{B}}{\partial t} = \nabla \times (\vec{u} \times \vec{B} - \eta \nabla \times \vec{B}) \quad (2)$$

(where η indicates the coefficient of Ohmic diffusivity) a dynamically-consistent velocity field $\vec{u}$ is found with inductive properties capable of sustaining $\vec{B}$ against dissipation. This is the *dynamo problem* of stellar plasmas, extensively discussed in many papers, e.g. in [9]. For our purposes, it is sufficient to remember the simple model developed, in a purely toroidal geometry, by [10], where the geometrical simplification permitted to derive exact $3D$ analytical solutions for magnetic buoyancy. From these solutions Trippella et al. [11] demonstrated that, at the bottom of a TDU episode, a penetration of protons can occur, with a mass distribution described by the relation:

$$\Delta M_p \simeq X_p \frac{4\pi\rho_E}{\alpha}[r_e^2 - \frac{2}{\alpha}r_e + \frac{2}{\alpha^2}] - [r_p^2 - \frac{2}{\alpha}r_p + \frac{2}{\alpha^2}]e^{-\alpha(r_e - r_p)} \quad (3)$$

Here, r_p and r_e are the values of the stellar radius at maximum penetration and at the envelope border, ρ_e is the average gas density at the convective border and X_p is the fractional mass of protons in the envelope. In the formula, the parameter α is adjusted so that proton pollution is limited to zones of viscosity small enough that the treatment by [10] of an almost ideal MHD holds. This depends on the stellar parameters of the case considered, but the polluted mass is normally around 0.004 – 0.005 $M_\odot$: see [12] for details.

The resulting proton abundance profile in the affected region is illustrated in Figure 1 (i). When H burning restarts in the shell above these layers, it produces the distribution of ^{13}C and of ^{14}N shown in Figure 1 (ii).

Nucleosynthesis models for Galactic stars, computed by assuming the presence, at each TDU episode, of a ^{13}C reservoir from equation (3), were presented recently by [12, 13]. When adopting nuclear inputs for neutron-capture cross sections and for weak interaction rates as discussed in [14] and after averaging over standard choices for the Initial Mass Function and for the Star Formation Rate, it is possible for such models to reproduce the solar abundance distribution of heavy elements quite well, obtaining a remarkable consistency with the complementary predictions of fast neutron captures (the r-process).

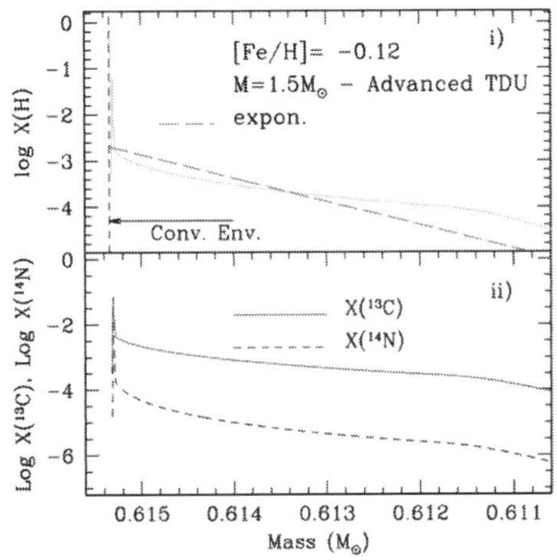

Figure 1. Upper panel: the reservoir (often called *pocket*) of protons mixed from the envelope, as a consequence of magnetic buoyancy computed according to [10, 11]. Lower panel: the ensuing distribution of ^{13}C and ^{14}N, established after burning is restarted in the H shell . As indicated in the label, the plot refers to an intermediate evolutionary stage of a 1.5 $M_{\odot}$ Thermall-Pulsing (*TP*) AGB star of almost solar metallicity

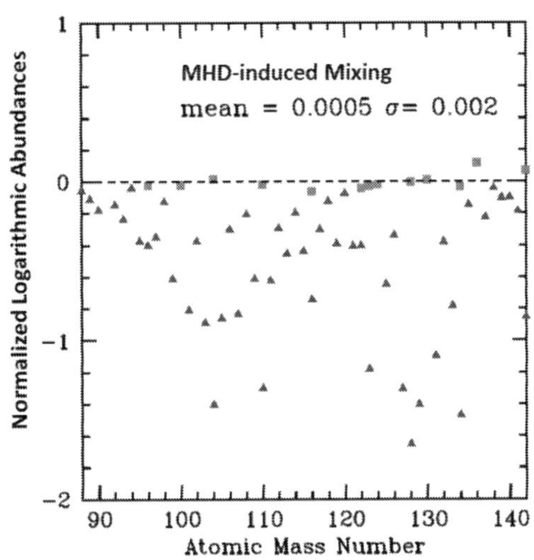

Figure 2. Predictions of the solar fractional abundance distribution from the models of stellar and galactic *s*-process nucleosynthesis discussed in the text, for nuclei from ^{88}Sr to ^{142}Ce.

This is shown in Figures 2 and 3.

Displayed in Figure 2 are the logarithms of solar abundance fractions due to *s*-processing, in the mass range $88 \lesssim A \lesssim 142$, resulting from the computations outlined above (see [15] for details). The nuclei shielded from fast decays (*s*-only isotopes) are indicated in red. As is seen in the figure, models for galactic *s*-process nucleosynthesis computed with ^{13}C pockets derived as in Figure 1 correctly predict for them abundance fractions close to unity. Blue triangles then represent the ensuing *s*-fractions of other nuclei, to be compared with expectations from the *r* process.

Such a comparison was performed by [15], by adopting the site-independent *waiting point* approach to fast neutron captures, with updated inputs for nuclear masses and decay rates. Figure 3 shows the discrepancies between the two sets of predictions. For the sake of the comparison, from *r*-process computations yielding expectations X_r for the nuclei in the given mass interval, the *s*-process *residuals*:

$$X_s = 1 - X_r \tag{4}$$

were derived. Then the ratios among the data of Figure 2 and these last estimates are shown in Figure 3. It is evident that a remarkably consistent view of neutron captures emerges, with few outlying nuclei. The peculiar nuclear problems affecting each of them was then discussed by [15] in the light of possible new measurements aimed at resolving these remaining discrepancies.

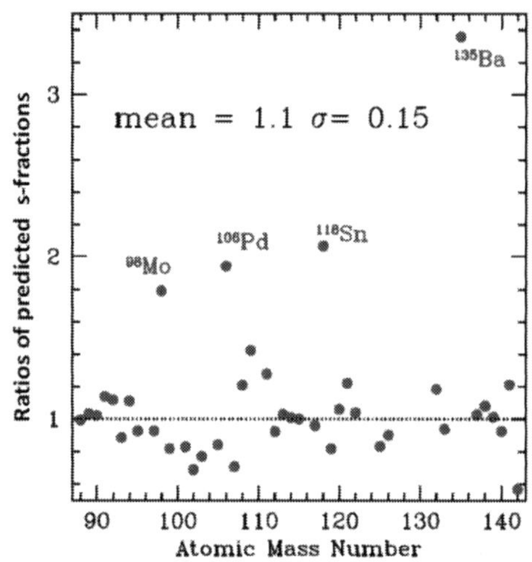

Figure 3. Discrepancies between the expectations of solar abundance fractions from slow neutron captures derived from the models discussed here and the *s*-residuals computed from *r*-process models according to relation (4).

The complementary results obtained by [14] in comparing *s*-process model predictions from the above scenario with isotopic ratios of heavy nuclei measured in presolar SiC grains of AGB origin [16] added further credibility to the global picture.

3 Characteristics of mixing models needed for AGB *s*-processing

In the proton distribution of Figure 1, panel i), the local concentration is very low (see the comparison, in the figure, with a purely exponential profile). This implies that, at the re-ignition of H-burning in the shell, proton captures do occur on ^{12}C producing ^{13}C, but minimal traces of ^{14}N can be synthesized (Figure 1, panel ii). Hence, neutrons made available by the $^{13}C(\alpha,n)^{16}O$ source remain largely unfiltered by intermediate-mass *poisons*, that would mainly result as daughters of ^{14}N-induced reactions.

These peculiarities of the models are crucial to obtain good agreement with various observations (see discussion in [11, 12, 15]). They induce remarkble differences with respect to previous models.

- First of all, the concentration of ^{19}F, produced from reactions starting at nitrogen, is strongly reduced with respect to what was obtained with exponential-like ^{13}C pockets. This fact favors agreement with fluorine observations [17].
- The extended ^{13}C pockets obtained in the above scenario make the nucleosynthesis of heavy nuclei from the ^{13}C(α,n)^{16}O source more efficient than before.
- A lower ^{14}N concentration also implies a reduced formation of ^{22}Ne, and a reduced efficiency of neutron captures in the warmer conditions of the thermal pulses, induced by the chain:

$$^{14}\mathrm{N}(\alpha,\gamma)^{18}\mathrm{F}(\beta^{+}\nu)^{18}\mathrm{O}(\alpha,\gamma)^{22}\mathrm{Ne}. \tag{5}$$

The available neutrons are therefore primarily released by the ^{13}C(α,n)^{16}O reaction, at very low neutron densities. The type of neutron captures descending from this situation implies in general less-pronounced branchings with respect to previous AGB models. This fact favors agreement with presolar grain isotopic ratios, as shown in [14].

All this ultimately descends from the fact that, in the models discussed, protons penetrate exponentially *in density* (equation (12) in [11]), not in mass. This introduces the quadratic terms in equation (3), implying an extended profile with a concentration of ^{13}C lower than in previous models (e.g. lower than in [3]).

As mentioned, the scenario depicted starts from the fact that the temperature gradient below TDU is not steep. Should temperature follow a profile less steep than $1/r$, this would produce *naturally* an expansion, as described by Parker in his seminal works on the solar wind [18, 19]. In [10] it was shown that, with a slightly steeper gradient of T, a non-static equilibrium forcing mixing is in any case induced by a magnetic field (hence a Lorentz force) varying in time.

This discussion implies therefore that any other hydrodynamical mechanism operating in the mild T gradient below TDU and introducing in the plasma forces varying in time, might be suited to induce mixing processes similar to the ones described here.

I strongly encourage stellar physicists to further verify this possibility, thus expanding our understanding of AGB mixing and of the neutron captures induced by it.

References

[1] S. Cristallo, *Europ. Phys. J., Web of Conf.* , in press (2022)
[2] M. Busso, R. Gallino, G.J. Wasserburg, *Ann. Rev. Astron. Astrophys.* **37**, 239 (1999)
[3] C. Arlandini, F. Käppeler, K. Wisshak, R. Gallino, M. Lugaro, M. Busso, O. Straniero, *Astrophys. J.* **525**, 886 (1999), astro-ph/9906266
[4] M. Schwarzschild, R. Harm, *Astrophys. J Suppl.* **73**, 35 (1968)
[5] R. Hōshi, *Progress of Theoretical Physics* **39**, *957 (1968)*
[6] R. Gallino, C. Arlandini, M. Busso, M. Lugaro, C. Travaglio, O. Straniero, A. Chieffi, M. Limongi, Astrophys. J. **497**, *388 (1998)*
[7] P. Charbonneau, Saas-Fee Advanced Course **39**, *187 (2013)*
[8] P. Hoyng, in Advances in Nonlinear Dynamics, *edited by A. Ferriz-Mas, M. Núñez (2003), pp. 1–36*
[9] P. Charbonneau, Living Reviews in Solar Physics **7**, *3 (2010)*
[10] M.C. Nucci, M. Busso, Astrophys. J. **787**, *141 (2014),* 1404.2503

[11] *O. Trippella, M. Busso, S. Palmerini, E. Maiorca, M.C. Nucci, Astrophys. J.***818**, *125 (2016),* 1512.06777

[12] *M. Busso, D. Vescovi, S. Palmerini, S. Cristallo, V. Antonuccio-Delogu, Astrophys. J.***908**, *55 (2021),* 2011.07469

[13] *D. Vescovi, S. Cristallo, M. Busso, N. Liu, Astrophys. J. Lett.***897**, *L25 (2020),* 2006.13729

[14] *S. Palmerini, M. Busso, D. Vescovi, E. Naselli, A. Pidatella, R. Mucciola, S. Cristallo, D. Mascali, A. Mengoni, S. Simonucci et al., Astrophys. J.***921**, *7 (2021),* 2107.12037

[15] *M. Busso, K.L. Kratz, S. Palmerini, W. Akram, V. Antonuccio-Delogu, Frontiers in Astronomy and Space Sciences* **in press** *(2022)*

[16] *N. Liu, S. Cristallo, D. Vescovi, Universe* **8**, *362 (2022),* 2206.13721

[17] *D. Vescovi, S. Cristallo, S. Palmerini, C. Abia, M. Busso, Astron. & Astrophys.***652**, *A100 (2021),* 2106.08241

[18] *E.N. Parker, Astrophys. J.***128**, *664 (1958)*

[19] *E.N. Parker, Astrophys. J.***132**, *175 (1960)*

Neutron captures in stellar nucleosynthesis

Sergio Cristallo[1,2,*]

[1]INAF - Osservatorio Astronomico d'Abruzzo, via Maggini snc, 64100, Teramo, Italy
[2]INFN - Sezione di Perugia, via A. Pascoli snc, 06126, Perugia, italy

Abstract. Apart from cosmological hydrogen and helium, chemical elements in the Universe are produced in stars, during both quiescent and explosive phases. The Sun chemical distribution witnesses the pollution from already extinct stellar generations at different epochs before the Solar System formation. The two major nucleosynthesis processes responsible for the formation of elements heavier than iron are the slow neutron capture process (the s-process) and the rapid neutron capture process (the r-process). A third, less common, nucleosynthesis channel is related to the intermediate neutron capture process (the i-process), whose existence is not ascertained yet. Finally, a few proton-rich isotopes are created by the p-process. I will show their characteristics and the stellar sites where they are at work.

1 Introduction

The chemical history of the Universe experienced a progressive enrichment of metals (i.e. elements with A≥12) at the expense of the cosmological hydrogen. The distribution of heavy elements (A>56) in the solar system shows an exponential decline, superimposed to a series of double-peaks, related to the s (slow) and the r (rapid) processes [1]. Elements heavier than iron can be synthesized mainly via neutron capture processes, whose occurrence is not hampered by the Coulomb barrier. This kind of nucleosynthesis may easily work at low energies, because neutron capture cross sections generally increase with decreasing energy.
The most common neutron fluxes in stars are either quite small (s-process: $n_n \sim 10^7$ cm^{-3}) or quite large (r-process: $n_n > 10^{20}$ cm^{-3}). In principle one might expect to encounter astrophysical neutron fluxes in the large region between these two densities. Although not universally recognized, those events are apparently not common, even if in the last 10 years there has been a growing evidence for the need of such intermediate neutron capture process (i-process; [2–4]). For the s-process, two stellar sites have been unequivocally identified: low mass Asymptotic Giant Branch (AGB) stars [5–7] and quiescent phases of massive star evolution [8]. On the other hand, r process is believed to occur during late evolutionary stages of massive stars (single or binary), the exact site being still matter of debate (Magneto-rotational SuperNovae [9], Neutron stars mergers (NSMs) [10], Collapsars [11]). In the following Sections, I will review the three aforementioned neutron capture processes.

*e-mail: sergio.cristallo@inaf.it

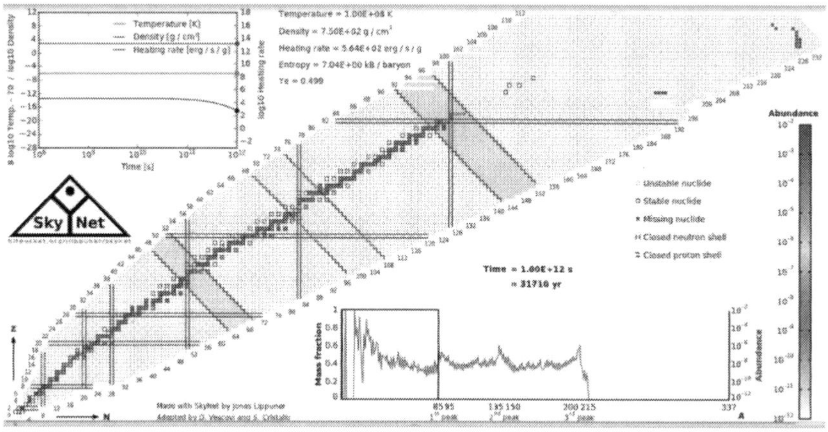

Figure 1. Typical pattern of the s-process. The plot has been produced with the PYTHON graphical interface of the SKYNET code [14].

2 The s-process

A typical s-process pattern is reported in Figure 1. The main s-process flow proceeds closely stick to the β stability valley (empty squares in the plot represent stable isotopes). Interestingly, there are isotope which are shielded by an r-process contribution by their stable isobars, named s-only, whose production is of paramount importance to calibrate s-process calculations (see [12] and references therein; see also [13] for a recent exploration of this subject). the main seed for the s-process is ^{56}Fe: this is due to its rather large abundance and neutron capture cross section. In Figure 1 mass fractions at the end of a typical s-process neutron exposure are reported. Three peaks clearly emerge from the distribution, corresponding to nuclei with neutron magic nuclei (N=50, N=82 and N=126). Those isotopes, characterized by closed shells configurations, are more stable with respect to their neighbors and, therefore, act as bottlenecks of the s-process fluency. The isotopes with the slowest neutron capture cross sections are ^{88}Sr, ^{138}Ba and ^{208}Pb, which marks the three s-process peaks. Another important quantity in s-process calculations is the neutron-to-seed ratio, defined as the ratio between the local neutron abundance and the local seed (mostly ^{56}Fe) abundance. The larger this ratio is, the more heaviest isotopes (as ^{208}Pb) are produced.

The two major neutron sources for the s-process are the ^{13}C(α,n)^{16}O and the ^{22}Ne(α,n)^{25}Mg reactions. The first reaction dominates in Thermally Pulsing low mass AGB stars (1 < M / M$_{\odot}$ < 4; [15, 16]) and it is responsible for the *main* component of the s-process (88≤A≤208). The AGB phase is attained by low and intermediate mass stars which are unable to activate advanced burnings in their core. During this phase, the stars consist of a partially degenerate C-O core, a radiative layers between an He-shell and an H-shell (where the s-process takes place) and by an expanded cool H-rich envelope. The products synthesized in the internal layers are carried to the surface by mixing episodes known as Third Dredge Ups (see [17] and references therein). Interested readers may download AGB surface abundances and yields from the FRUITY database[1] [6].

The other neutron source, the ^{22}Ne(α,n)^{25}Mg, is the main neutron producer in massive stars, accounting for the *weak* component of the s-process (A<88). As a matter of fact, elements between iron and the first s-process peak are efficiently synthesized by the weak s-process [8],

[1] http://fruity.oa-teramo.inaf.it/

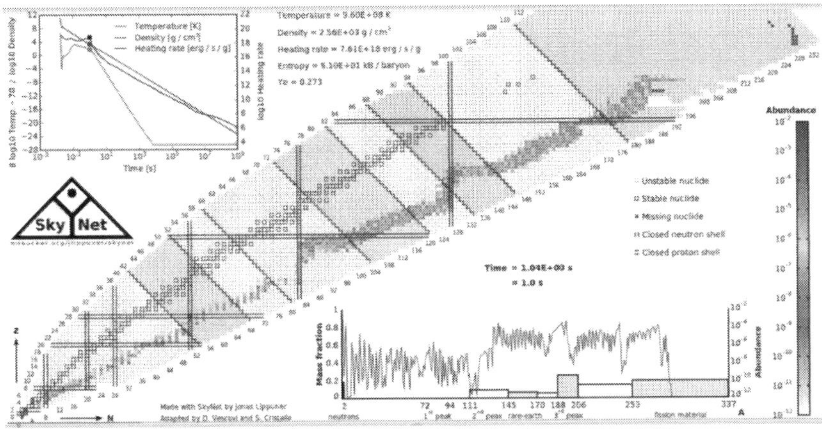

Figure 2. Typical pattern of the r-process. The plot has been produced with the PYTHON graphical interface of the SKYNET code [14].

which is at work during the core He-burning (n~ 10^6 cm^{-3}) and the C-shell burning (n~ 10^{12} cm^{-3}) phases of massive (M> 9 $M_\odot$) stars evolution.

3 The r-process

The remaining half of the heavy elements in the Universe are synthesized via the r-process. The latter has been proved to be at work during NSMs, by observing and interpreting the electromagnetic transient (the so-called kilonova) following the gravitational event GW170817. Since then, an exponentially growing number of studies on this subject have become available, making the last years an exciting period for the investigation of the r-process nucleosynthesis. The electromagnetic counterpart (AT2017gfo) of the gravitational wave detection GW170817 provided the first record of the in-situ operation of the r-process. During this nucleosynthesis process, extremely large neutron densities allow the production of neutron-rich isotopes far from the β-stability valley. The physics of the r-process is by far more complex with respect to the s-process. As a matter of fact, due to the difficulties in modelling explosions and because of the large uncertainties in nuclear inputs, the r-process contribution to the solar distribution is estimated by subtracting that coming from the s-process, i.e. r=1-s [13, 18].

In Figure 2 we report the typical r-process path attained in a NSM 1 second after the merger. For many key quantities only theoretical estimates are currently available, making predictions for the r-process less robust than the corresponding ones related to the s-process. The r-process is normally characterized by a huge amount of free neutrons. Thus, due to the large neutron-to-seed ratios (up to several 1000), the nucleosynthesis becomes insensitive to the initial composition and the material loses memory of the exact thermodynamic conditions at ejection. However, there is a late phase, in which neutron captures and decays operate on very similar time scales (r-process freeze-out). During this phase, neutron captures may shape the final r-process distribution. Improvements are expected from an in-depth knowledge of properties of the nuclei far from stability, from both the theoretical and experimental point of view. To that purpose, sensitivity studies aiming at identifying the most important nuclear properties to be measured in order to reduce the uncertainty related to the r-process are particularly helpful (see e.g. [19]). Sensitivity studies play a key role in facilitating state-of-the-art measurements as they provide crucial astrophysical motivation to focus experimental campaigns

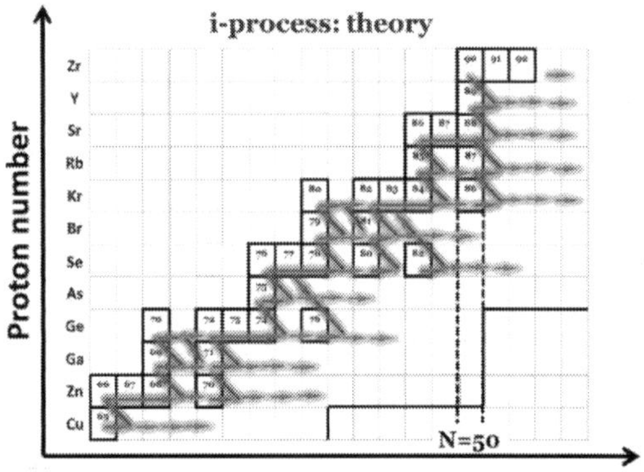

Figure 3. Typical pattern of the i-process. Green arrows mark the departure from the β stability valley. Red arrows indicate the main s-process path.

on the most impactful nuclei. However, the results of any sensitivity study have to be taken with a grain of salt, because even considering just one stellar site (as NMSs), the physical characteristics of various tracers at different angles can lead to completely different chemical patterns. The latter are mainly regulated by the neutronization level (Y_e, or electron fraction). Finally, details about interactions with neutrinos are fundamental. In fact, the neutrino transport determines the local Y_e, shaping the final r-process distribution. Besides the many stellar sites proposed to host the r-process, three are the major candidates:

- Neutron Star Mergers: these events are the only stellar site in which r-process has been demonstrated to be at work (see e.g. [20]). A rich nucleosynthesis is expected, involving also elements lighter than lantanides (as Sr, see [21]), but not H or He (see [22]);
- Magneto-rotational Supernovae: while standard supernovae are not able to develop a full r-process (but may instead be responsible for the weak component of the r-process, [23]), if the effects of strong magnetic fields are included, a complete r-process takes places [9]
- Collapsars: accretion disks around Black holes are the third candidate to host the r-process [11]. In such a case, also lighter isotopes (i.f. those belonging to the iron peak) can be produced.

To date is extremely challenging to disentangle among these three scenarios. For this reason Galactic Chemical Evolution models are required to weight their contributions and eventually identify the most probable candidates to host the r-process [24].

4 The i-process

The intermediate neutron capture process (i-process) was first theorized by [25]. It develops whenever protons are mixed in regions with typical He-burning temperatures (T ~ 200-300 MK). In this situation, hydrogen burns on-fly producing ^{13}N and/or ^{13}C, on a timescale of the order of hours (thus intermediate between tens of thousands year of the s-process and 1 s of the r-process). In such a situation, a few (from 4 to 6-7) neutron captures may occur on stable isotopes (see Figure 3). Then, when the neutron exposure comes to an end, unstable isotopes decay back to their stable isobars. It is still a matter of debate if such an i-process has sizable effects or not on the Galactic Chemical Evolution history. There are examples whose chemical patterns can be fitted very well by this process. The best known example are the so-called Carbon Enhanced Metal Poor (CEMP) s/r stars, named in such a way due to their peculiar chemical surface distributions. To date, two are the major candidates to host the i-process: Low-Metallicity Low-Mass AGB stars [2, 4, 5] and Rapidly Accreting White Dwarfs [3].

References

[1] E.M. Burbidge, G.R. Burbidge, W.A. Fowler, F. Hoyle, Reviews of Modern Physics **29**, 547 (1957)
[2] S. Cristallo, D. Karinkuzhi, A. Goswami, L. Piersanti, D. Gobrecht, ApJ **833**, 181 (2016), 1610.05475
[3] P.A. Denissenkov, F. Herwig, P. Woodward, R. Andrassy, M. Pignatari, S. Jones, MNRAS **488**, 4258 (2019), 1809.03666
[4] A. Choplin, L. Siess, S. Goriely, A&A **648**, A119 (2021), 2102.08840
[5] S. Cristallo, O. Straniero, R. Gallino, L. Piersanti, I. Domínguez, M.T. Lederer, ApJ **696**, 797 (2009), 0902.0243
[6] S. Cristallo, L. Piersanti, O. Straniero, R. Gallino, I. Domínguez, C. Abia, G. Di Rico, M. Quintini, S. Bisterzo, ApJS **197**, 17 (2011), 1109.1176
[7] S. Cristallo, O. Straniero, L. Piersanti, D. Gobrecht, ApJS **219**, 40 (2015), 1507.07338
[8] M. Pignatari, R. Gallino, M. Heil, M. Wiescher, F. Käppeler, F. Herwig, S. Bisterzo, ApJ **710**, 1557 (2010)
[9] M. Reichert, M. Obergaulinger, M. Eichler, M.Á. Aloy, A. Arcones, MNRAS **501**, 5733 (2021), 2010.02227
[10] D. Kasen, B. Metzger, J. Barnes, E. Quataert, E. Ramirez-Ruiz, Nature **551**, 80 (2017), 1710.05463
[11] D.M. Siegel, J. Barnes, B.D. Metzger, Nature **569**, 241 (2019), 1810.00098
[12] S. Cristallo, C. Abia, O. Straniero, L. Piersanti, ApJ **801**, 53 (2015), 1501.00544
[13] M. Busso, W.K. Kratz, S. Palmerini, W. Akram, V. Antonuccio-Deloglu, Frontiers in Astronomy and Space Sciences, in press (2022)
[14] J. Lippuner, L.F. Roberts, ApJ **815**, 82 (2015), 1508.03133
[15] R. Gallino, C. Arlandini, M. Busso, M. Lugaro, C. Travaglio, O. Straniero, A. Chieffi, M. Limongi, ApJ **497**, 388 (1998)
[16] M. Busso, R. Gallino, G.J. Wasserburg, ARA&A **37**, 239 (1999)
[17] O. Straniero, R. Gallino, S. Cristallo, Nuclear Physics A **777**, 311 (2006), astro-ph/0501405
[18] N. Prantzos, C. Abia, S. Cristallo, M. Limongi, A. Chieffi, MNRAS **491**, 1832 (2020), 1911.02545
[19] D. Vescovi, R. Reifarth, S. Cristallo, A. Couture, Frontiers in Astronomy (2022)
[20] E. Pian, P. D'Avanzo, S. Benetti, M. Branchesi, E. Brocato, S. Campana, E. Cappellaro, S. Covino, V. D'Elia, J.P.U. Fynbo et al., Nature **551**, 67 (2017), 1710.05858
[21] D. Watson, C.J. Hansen, J. Selsing, A. Koch, D.B. Malesani, A.C. Andersen, J.P.U. Fynbo, A. Arcones, A. Bauswein, S. Covino et al., Nature **574**, 497 (2019), 1910.10510
[22] A. Perego, D. Vescovi, A. Fiore, L. Chiesa, C. Vogl, S. Benetti, S. Bernuzzi, M. Branchesi, E. Cappellaro, S. Cristallo et al., ApJ **925**, 22 (2022), 2009.08988
[23] S. Goriely, A. Bauswein, O. Just, E. Pllumbi, H.T. Janka, MNRAS **452**, 3894 (2015), 1504.04377
[24] M. Van der Swaelmen, C. Viscasillas Vázquez, G. Cescutti, L. Magrini, S. Cristallo, D. Vescovi, S. Randich, G. Tautvaišienė, V. Bagdonas, T. Bensby et al., arXiv e-prints arXiv:2207.14747 (2022), 2207.14747
[25] J.J. Cowan, W.K. Rose, ApJ **212**, 149 (1977)

Influence of crystal structures on electron screening

Aleksandra Cvetinović[1,⋆], *Dijana* Djeordjić[2,3], *Giovani Luca* Guardo[4,5], *Mitja* Kelemen[1,2], *Marco* La Cognata[4], *Livio* Lamia[4,5,6], *Sabina* Markelj[1], *Rosario Gianluca* Pizzone[4], *Thomas* Schwarz-Selinger[7], *Isabela* Tišma[8], *Matjaž* Vencelj[1], *Jelena* Vesić[1], and *Matej* Lipoglavšek[1]

[1] *Jožef Stefan Institute, Ljubljana, Slovenia*
[2] *Jožef Stefan International Postgraduate School, Ljubljana, Slovenia*
[3] *University of Banja Luka, Faculty of Mechanical Engineering, Banja Luka, Bosnia and Herzegovina*
[4] *INFN-Laboratori Nazionali del Sud, Catania, Italy*
[5] *Dipartimento di Fisica e Astronomia "E. Majorana", University of Catania, Catania, Italy*
[6] *Centro di Fisica Nucleare e Struttura della Materia, University of Catania, Catania, Italy*
[7] *Max Planck Institute for Plasma Physics, Garching, Germany*
[8] *Rudjer Bošković Institute, Zagreb, Croatia*

Abstract. For nucleosynthesis calculations, precise reaction rates should be known at energies within the Gamow window. At these energies, electron screening cannot be neglected. Despite the significance of the effect, a huge disagreement between experimental data and theoretical predictions is still not understood. In order to address to this problem, we investigated the dependence of the electron screening potential on the target host lattice structure by measuring the rate of the $^{2}H(^{19}F,p)^{20}F$ reaction in zirconium, titanium and palladium targets containing deuterium .

1 Introduction

Accurate measurements of nuclear reactions between charged particles show an unexpectedly large enhancement of the cross section at energies within the Gamow window [1]. The enhancement is attributed to the presence of atomic electrons that screen the nuclear charge, effectively reducing the repulsive Coulomb barrier between interacting particles. Consequentially, the probability for the nuclear reaction is increased. The screening effect dominates reaction rates at thermonuclear energies, but at higher energies its contribution can be neglected. Moreover, experimental results (see Refs. [2–9] and references therein) unequivocally indicate the incorrectness of available models, since the amplitudes of measured screening potentials (U_e) strongly exceed the theoretically predicted values [10]. While the theory predicts an electron screening potential independent of the target host material, measurements report a strong dependence of the cross section enhancement on the target host and the metallurgy of the solid lattice [9]. Namely, when gaseous targets are used, the electron screening potentials remain within the adiabatic limit. However, the cross sections are especially enhanced in cases when the target nuclei are implanted into a solid lattice, often more than an order of magnitude above the predictions. Therefore, the problem remains and has to be solved, before one can apply electron

⋆e-mail: aleksandra.cvetinovic@ijs.si

screening to the nucleosynthesis calculations and try to understand what would be the consequences of electron screening in the thermonuclear processes in stellar plasma.

A simplified static theoretical approach [10] is usually employed for electron screening calculations, which assumes that the atomic electrons form a uniformly charged spherical shell, with an atomic radius R_a, around target nuclei. The reaction cross section $\sigma(E)$ is then enlarged by an enhancement factor f defined as the ratio of the screened and bare-nucleus cross sections:

$$f = \frac{\sigma(E + U_e)}{\sigma(E)}, \tag{1}$$

where E represents the energy in the center of mass system and U_e the electron screening potential that can be simply calculated as:

$$U_e = \frac{Z_1 Z_2 e^2}{4\pi\epsilon_0 R_a}, \tag{2}$$

where Z_1 and Z_2 are the charge numbers of the projectile and target nuclei. This result is called the adiabatic limit [10] and represents the maximum value for the screening potential. To simplify the analysis, when nuclear reactions are studied at low energies, the cross section σ is usually defined using the astrophysical $S(E)$-factor, which in the case of non-resonant reactions varies smoothly with energy [1]:

$$\sigma(E) = \frac{S(E)}{E} e^{-2\pi\eta}. \tag{3}$$

Here η is the Sommerfeld parameter.

2 Experiment

The experimental study of the electron screening effect was performed using the 2 MV Tandetron accelerator at the Microanalytical Center of Jožef Stefan Institute (JSI). We investigated the dependence of the electron screening potential on the target host lattice structure by measuring the rate of the $^{2}H(^{19}F,p)^{20}F$ reaction in thick zirconium, two thick and one thin titanium and two different thick palladium deuterium-containing targets. Based on our previous findings [9] suggesting that the preparation of the host material and location of the target nuclei in the metallic lattice can influence the electron screening, our goal was to find different values of U_e in at least two targets and then to understand which parameters of those targets differ and cause high electron screening. Our latest results are presented below.

2.1 Target preparation

For our study of electron screening, we chose Zr, Ti and Pd as hosting materials because these metals have the ability to absorb large volumetric quantities of deuterium. The Zr target was a 0.25 mm thick 99.8 % pure foil purchased at ChemPUR. The first Ti target was a 1 mm thick 99 % pure foil purchased at Goodfellow. The second titanium target was made by pressing the TiD powder to a thickness of 1 mm into a cylindrical hole (14 mm diameter and 1 mm depth) in 2 mm thick Cu backing. The third Ti target was a 330 nm thick Ti evaporated on Ta backing. The first Pd target was 100 μm thick soft Pd foil purchased at ChemPUR and the second one was a 100 μm thick hard Pd foil produced according to our specifications at Zlatarna Celje. It was cold rolled from a thickness of about 2.5 mm to 0.1 mm. This foil was much less flexible than Chempur's one. Purities of both foils were above 99.9%. In order to prepare the targets for electron screening studies, Zr and Ti foils were implanted using Tectra IonEtch ion gun. The foils have been bombarded for about 24 h with the deuterium beam accelerated by an extraction voltage of 3.5 kV. The beam current was 400 μA. During implantation, the targets were mounted on a massive copper holder which effectively prevented substantial heating of the foils. Due to the deuterium diffusion, foils had high deuterium concentrations at depths much higher than the implantation depth at 3.5 keV (41 nm [11]). Pd foils were loaded by leaving the palladium in deuterium gas at the pressure of 1 bar and temperature of 24°C for 24 hours. In the soft Pd foil, maximum concentrations of 70% of D per metallic atom were achieved. This is consistent with the limit of hydrogen absorption at normal pressures [12]. However, the hard Pd foil could be loaded only up to 47% of D per metallic atom.

The deuterium depth distribution in the targets was determined with the Nuclear Reaction Analysis (NRA) technique [13]. For this purpose, high energy protons emitted in the $^2H(^3He,p)^4He$ reaction were measured at six ^3He energies, in an energy region from 0.629 to 4.297 MeV. The deuterium depth profiles were obtained by fitting the NRA spectra using the SIMNRA code [14]. Fig. 1 shows deuterium depth distributions measured in our targets after the ^{19}F+d experiment. Deuterium concentration errors are estimated to be ~4%. Note that the 9.2 MeV ^{19}F ion beam is fully stopped in Pd at the depth of 2.62 μm, in Zr at the depth of 3.91 μm and in Ti at the depth of 4.07 μm.

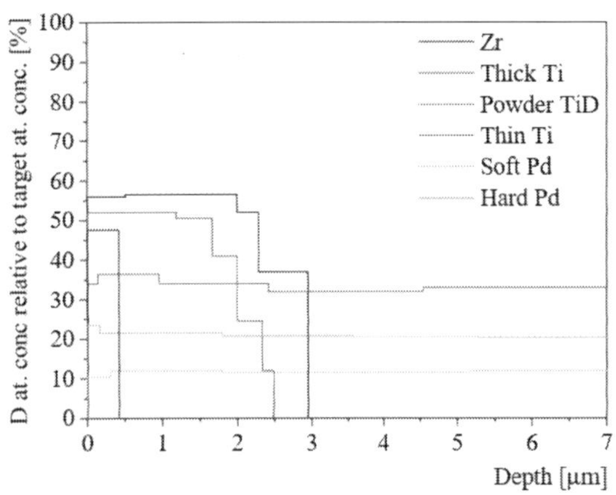

Figure 1. Deuterium concentration relative to the number of target atoms as a function of depth in the target, measured in Zr (black line), thick Ti (red line), powder Ti (green line), thin Ti (blue line), soft Pd (purple line) and hard Pd (light blue line) targets using the Nuclear Reaction Analysis technique. Deuterium concentration errors are estimated to be ~4%.

2.2 Experimental setup

The reaction between ^2H and ^{19}F produces the radioactive ^{20}F which decays with a half-life of 11 s through β- decay to the stable ^{20}Ne. ^{20}Ne de-excites to the ground state by emitting a γ-ray with an energy of 1634 keV. The branching ratio for this transition is 99.9% [15]. To determine the screened cross section, yield of these gammas was measured using a High-Purity Germanium (HPGe) detector positioned 57 mm from the target at an angle of 135° with respect to the ^{19}F ion beam direction. The intrinsic detector efficiency was 53% relative to a 3" x 3" NaI detector. It had an efficiency of 0.6% and a resolution of 2.2 keV at the 1.3 MeV ^{60}Co peak. The ^{19}F ion beam with currents of about 0.5 μA was used to study the ^{19}F+d reaction in an energy range between 3.089 and 9.200 MeV. The numbers of incident ions were deduced by measuring the charge collected on the electrically isolated target chamber.

In order to effectively prevent substantial heating of the targets during the experiment, foils were mounted on a massive copper holder. Targets were positioned in a high vacuum chamber with its surface orientated perpendicularly to the beam direction. During the experiment, we monitored deuterium loss in all targets by repeatedly measuring yields at the beam energy of 7.671 MeV before and after each measurement at other energies. Almost negligible changes in deuterium concentration were detected in Zr and Ti targets. However, Pd targets showed a loss of deuterium during all measurements. In order to correct for the deuterium loss, we normalized the detected yields for a given beam energy to the average of the two control measurements.

3 Results

In the case of the thin Ti target, the reaction cross section σ was calculated from [16]:

$$N_\gamma = \sigma \epsilon W_\gamma N_F n_D \frac{\rho N_A x}{M}. \tag{4}$$

Here ϵ is the efficiency of the detector, W_γ is the angular distribution factor for emitted γ-rays (W_γ=1 since the γ-ray angular distribution is isotropic after β decay [15]), N_F is the number of incident ^{19}F ions, N_γ is the experimentally measured number of γ rays and the value $n_D \frac{\rho N_A d}{M}$ represents the surface density of deuterium atoms in the target (here n_D is the number of deuterium atoms per crystal lattice atom, x is the target thickness, N_A, ρ and M are the Avogadro's number, target density and molar mass).

In the case of thick targets, Eq. 4 has to be transformed into a differential form and integrated over energies from the beam energy E_0 to E_x:

$$N_\gamma = N_F n_D \frac{\rho N_A}{M} \int_{E_0}^{E_x} \epsilon W_\gamma \frac{\sigma_E}{dE_F/dx} dE_F, \tag{5}$$

where E_x is the remaining energy of the beam after passing trough the target layer implanted with D. The stopping power dE_F/dx was calculated using the SRIM code [11].

Since there is no available cross section for the ^{19}F+d reaction in the studied energy region, to determine the bare-nucleus cross section, we measured the same reaction, but this time in forward kinematics, in which we never observed a large electron screening effect, except for the p+d reaction [17, 18]. The measurement was performed with the 3MV accelerator at Max Planck Institute for Plasma Physics in Garching in an energy range from E_d=(303 to 998 keV) on thin CaF_2 targets positioned perpendicularly to the beam. The 1634 keV γ-ray yield was measured using the same HPGe detector as was used for the experiments in inverse kinematics, but this time positioned at an angle of 0° with respect to the ion beam direction, 4.82 mm from the target. More details on the used experimental technique and the cross section data are in [19]. In this experiment, by fitting the experimental data we obtained the bare-nucleus astrophysical S-factor (center-of-mass energy E is given in MeV.):

$$S(E)_{^{19}F+d} = 19380 - 4596E - 3218E^2\,[MeVb], \tag{6}$$

that was used to calculate enhancement factors and electron screening potentials for the ^{2}H(^{19}F,p)^{20}F reaction in all our targets. In the upper panel of Fig. 2 we show the integrated enhancement factors as a function of fluorine beam energy in the center-of-mass system that we obtained in three Ti Targets, while integrated enhancement factors obtained in Zr and two Pd targets are shown in the lower panel. The obtained screening potentials for each target are listed in Table 1.

4 Conclusions

We studied the electron screening effect in the ^{2}H(^{19}F,p)^{20}F nuclear reaction on Zr, three different Ti and two different Pd targets containing deuterium. In all targets we measured different values of the

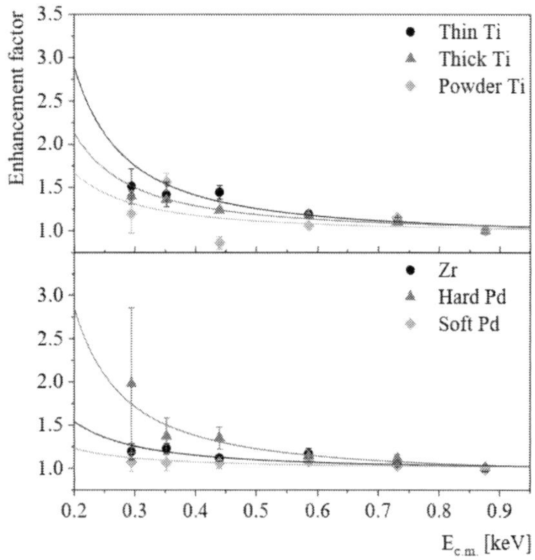

Figure 2. Integrated enhancement factors as a function of the fluorine beam energy in the center-of-mass system for the $^{2}H(^{19}F,p)^{20}F$ reaction obtained in three Ti targets (upper panel), and in the Zr and two Pd targets (lower panel). Points represent experimental data and the solid lines represent least-squares fits to the data.

Table 1. The electron screening potentials U_e measured in the $^{2}H(^{19}F,p)^{20}F$ reaction in six different targets in comparison with the predicted screening potential given by the adiabatic model.

Target	U_e [keV]
Zr	7.0±1.9
Soft Pd	3.2±1.9
Hard Pd	18.2±3.3
Thin Ti	18.0±4.9
Thick Ti	12.3±2.8
Powder Ti	8.4±9.2
U_{ad} [keV]	2.19

electron screening potentials. In our powder Ti target we did not detect electron screening different from zero, within the error bars (U_e=8.4±9.2 keV). In the soft Pd target the measured screening potential U_e=3.2±1.9 was in agreement with the theoretical value (U_{ad}=2.19 keV). In the remaining four targets we measured high electron screening potentials, that were up to an order of magnitude above the theoretical model. Namely, in Zr target we measured electron screening potential to be U_e=7.0±1.9, in the thick Ti foil U_e=12.3±2.8 and in the thin Ti and Hard Pd targets we measured the highest screening of U_e=18.0±4.9 and U_e=18.2±3.3, respectively. Since in each target we measured a different screening potential for the same nuclear reaction, which is contrary to the predictions given by the available theoretical model, we found that the screening effect is not linked to the static electron densities around interacting nuclei and that probably, a dynamic approach should be applied. Our findings clearly show that electron screening is strongly linked to the host's crystal lattice structure and electron densities at locations of the target nuclei in the metallic lattice. However, in order to understand this link, additional investigations are required. Our future plans are to apply quantitative methods, such as neutron and X-ray diffraction and nuclear magnetic resonance analysis, in order to test this link.

References

[1] C. Iliadis, *Nuclear Physics of Stars - Second, Revised and Enlarged Edition* (Wiley-VCH, Weinheim, 2015)
[2] K. Czerski et al., Europhys. Lett. **68**, 363 (2004)
[3] J. Kasagi et al., J. Phys. Soc. Jpn. **73**, 608 (2004)
[4] F. Raiola et al., J. Phys., G **31**, 1141 (2005)
[5] J. Cruz et al., Phys. Lett., B **624**, 181 (2005)
[6] T. S. Wang et al., J. Phys. G: Nucl. Part. Phys. **34**, 2255 (2007)
[7] A. Huke et al., Phys. Rev. C **78**, 015803 (2008)
[8] M. Lipoglavšek et al., Eur. Phys. J., A **44**, 71 (2010)
[9] A. Cvetinović et al., Phys. Rev., C **92**, 065801 (2015)
[10] H. J. Assenbaum et al., Z. Phys. A: At. Nucl. **327**, 461 (1987)
[11] J. F. Ziegler et al., *The Stopping and Range of Ions in Matter* (Lulu Press Co., Morrisville, NC, 2008) www.srim.org
[12] G. Alefeld and J. Völkl, *Hydrogen in metals II, Application-oriented properties* (1978) 81
[13] P. Petersson, et al., Nucl. Instrum. Methods Phys., B 268, 1838 (2010)
[14] M. Mayer, *SIMNRA User's Guide* (Report IPP 9/113, Max-Planck-Institut für Plasmaphysik, Garching, Germany, 1997) https://mam.home.ipp.mpg.de
[15] D. R. Tilley et al., Nucl. Phys. A **636**, 249 (1998)
[16] C. Rolfs and R. W. Kavanagh, Nucl. Phys. A **455**, 179 (1986)
[17] M. Lipoglavšek et al., Phys. Lett., B **773**, 553 (2017)
[18] J. Vesić et al., Eur. Phys. J., A **50**, 153 (2014)
[19] A. Cvetinović et al., submitted to Phys. Lett., B.

ANC method: experimental approach and recent results

G. D'Agata[1,2,3], A. I. Kilic[3], V. Burjan[3], A. Cassisa[3], J. Mrázek[3], and V. Glagolev[3]

[1]INFN - Laboratori Nazionali del Sud, Via Santa Sofia 62, 95123, Catania, Italy
[2]Dipartimento di Fisica e Astronomia, Università degli Studi di Catania, Via S. Sofia, 64, 95123, Catania, Italy
[3]Nuclear Physics Institute of the Czech Academy of Sciences, 250 68 Řež, Czech Republic

Abstract. The Asymptotic Normalization Coefficient (ANC) method, has proven to be useful in retrieving the direct part of the radiative capture cross section for a number of reactions of astrophysical interest. In this work, some of the latest results obtained by the AsFin2 group of the LSN-INFN (Catania) and the OJR of the NPI (Řež) will be briefly discussed.

1 Introduction

One of the biggest problems for nuclear astrophysics has always been to experimentally retrieve the cross section at astrophysical energies: the presence of the Coulomb barrier strongly decreases the overall cross section in the range of interest (usually between some keV and some MeV). Also, many reactions involve radioactive nuclei with a small half-life in the entrance channel, making a direct measurement sometimes almost unfeasible. For this reason, many indirect methods have been developed in the last decades, and among these the Asymptotic Normalization Coefficient (ANC)
textslciteMuc1,Xu one is useful to extract the direct capture cross section for (p,γ) [3], (n,γ) [4] and (α,γ) [5] reactions of astrophysical interest from a suitable one-particle transfer. In the case of highly unstable nuclei with small half-lifes, an extention to the method has been developed involving mirror reactions [6, 7]. In the following work the ANC method will be presented, and recent results regarding the study of the ^{26}Si$(p,\gamma)^{27}$P (performed using the mirror nuclei procedure) [8], the ^{6}Li$(p,\gamma)^{7}$Be [9] and ^{3}He$(\alpha,\gamma)^{7}$Be [10] reactions will be discussed.

2 The ANC method

As stated in the introduction, the ANC method allows to indirectly study direct capture reactions of astrophysical interest from one-particle transfer reactions. In the general case, a direct transfer reaction can be visualized as a $X + A \rightarrow Y + B$ process, where both the X and B nuclei can be considered as $X = Y + a$ and $B = A + a$, a being the transferred particle.
Using the Distorted Wave Born Approximation (DWBA), the matrix element for the transition between the initial and final state can be written as

$$M = \left\langle \chi_f^{(-)} I_{Aa}^{B} \left|\Delta V\right| I_{Ya}^{X} \chi_i^{(+)} \right\rangle \quad (1)$$

where I^B_{Aa} and I^X_{Ya} are the radial overlap integrals of the two systems $Y + a$ and $A + a$ with the nuclei X and B respectively. The quantities $\chi_i^{(+)}$ and $\chi_f^{(-)}$ are the distorted wave-functions for the entrance and exit channels, while ΔV is the perturbative part of the total potential, composed by a real (V) and an optical (U) part ($\Delta V = V - U$). For a general nuclear process $a + b \to c$, with $c = a + b$, both the overlap functions in equation 1 can be written using the so-called spectroscopic factor (SF) – in general S_{ab} – that has been widely used in the last fifty years to estimate the occupation probability of the orbitals:

$$I^c_{ab,l_c,j_c}(r_{ab}) = S^{1/2}_{ab,l_c,j_c}\phi_{n_c,l_c,j_c}(r_{ab}) \tag{2}$$

$\phi_{n_c l_c j_c}(r_{ab})$ being the bound-state wave function [12].
The description of the cross section using the SF depends on the binding portential and optical model parameters (OMP) [12] used to describe the reaction: the first one is in fact adjusted according to the binding energy to reproduce the bound state, while the OMP are usually extracted from elastic scattering. The SF is sensitive to the selection of the optical model adopted, and different families of potentials can reproduce the angular distributions retrieved from experimental data [14]. Using the DWBA it is possible to reproduce the experimental angular distribution in the following way:

$$\frac{d\sigma^{exp}}{d\Omega} = \sum_{j_B, j_X} S_{Aa,l_B,j_B} S_{Ya,l_X j_X} \sigma^{DW}_{l_B,j_b,l_X,j_X} \tag{3}$$

Here $\frac{d\sigma^{exp}}{d\Omega}$ and $\sigma^{DW}_{l_B j_b l_X j_X}$ are related by the two SF for the entrance and exit channel: while one of those is usually known, the other is left to be determined.
Mukhamedzhanov and Timofeyuk [11] pointed out that the direct part of the radiative cross section at low energies contains the same radial overlap integral (I^c_{ab,l_c,l_j}) of a direct transfer reaction: this happens because such capture occurs at large distances from the nucleus, where the behaviour can be called asymptotic. The asymptotic behaviour can be therefore used to extract the cross section of reactions of astrophysical interest: in this case is possible to write both the radial overlap functions and bound-state wave functions as

$$I^B_{Aa,l_c,j_c}(r_{ab}) \xrightarrow{r_{ab}>R_n} C^c_{ab,l_b,j_b} \frac{W_{-\eta,l_c+\frac{1}{2}}(2k_{ab}r_{ab})}{r_{ab}} \tag{4}$$

and

$$\phi_{ab,l_c,j_c}(r_{ab}) \xrightarrow{r_{ab}>R_n} b^c_{ab,l_c,j_c} \frac{W_{-\eta,l_c+\frac{1}{2}}(2k_{ab}r_{ab})}{r_{ab}} \tag{5}$$

In both equations (4) and (5) $W_{-\eta,l_c+\frac{1}{2}}$ is the Whittaker function, which asymptotic behaviour can be written as $W_{-\eta,l_c+\frac{1}{2}}(-2k_I r) \to e^{-k_I r + \eta_I ln(2k_I)}$, while the coefficients b and C are the so-called single-particle ANC (SPANC) and ANC, respectively. By substitution of equations (4) and (5) in (3), we obtain:

$$\frac{d\sigma}{d\Omega} = \sum_{j_B, j_X} \left(C^B_{Aa,l_B,j_B}\right)^2 \left(C^X_{Ya,l_X,j_X}\right)^2 \frac{\sigma^{DWBA}_{l_B,j_B,l_X,j_X}}{b^2_{Aa,l_B,j_B} b^2_{Ya,l_X,j_X}} \tag{6}$$

Using DWBA, we are finally able to extract the ANC's for the direct part of the cross section for reactions of astrophysical interest. This approach is useful because the ANC's have a small dependence on the chosen nucleon binding potential [12], making the cross section nearly independent from b^2. This last statement is not true in the case of tightly bound

core+particle systems: the sensitivity to the potential in the continuum state in fact increases with the separation energies, and the zero-energy S-factor is not uniquely determined by the experimentally determined ground state ANC [13].
More recently [6], an extension to the ANC method has been proposed involving the so-called mirror nuclei. Such a tool has proven to be useful in the case of proton (p) or neutron (n) transfer involving strongly unstable nuclei [6–8]: being $A + p \to B$ the proton transfer, the proton ANC (C_p^{A+p}) can in fact be extracted from the neutron ANC (C_n^{D+n}) of a suitable mirror partner $D+n \to E$, where D and E have inverted number of protons and neutrons with respect to A and B, respectively. The two ANC share the relation

$$\left(C_p^{A+p}\right)^2 = R_{mirr}\left(C^{C+n}\right)^2 \tag{7}$$

where R_{mirr} can be calculated as

$$R_{mirr} = \left|\frac{F_l\left(ik_pR_N\right)}{k_pR_nj_l\left(ik_nR_N\right)}\right|^2 \tag{8}$$

F_l being the regular Coulomb wave function, j_l the Bessel function of l–th order, and R_N the radius of the nuclear interior. The quantities k_p and k_n are related to the neutron and proton separation energy via the relation $k = \sqrt{\frac{2\mu\epsilon}{\hbar^2}}$ [6, 7].
In the case of the population of an excited state near the treshold, the mirror state will exist as a resonance. Thus a relation exists that ties the C^2 of the direct capture in the excited state and the Γ of the unbound state in the mirror counterpart [6]:

$$\frac{\Gamma_{p(n)}}{\left|C_{n(p)}\right|^2} = R_\Gamma \approx R_0^{res} = \frac{\hbar^2k_{p(n)}}{\mu}\left|\frac{F_l\left(ik_{p(n)}R_N\right)}{k_{p(n)}R_Nj_l\left(ik_{n(p)}R_N\right)}\right|^2 \tag{9}$$

3 ANC application

3.1 The ^{26}Si$(p,\gamma)^{27}$P reaction

^{26}Si production and destruction channels are important in the context of the long-standing problem of ^{26}Al ($T_{1/2}$ = 0.72 Myrs). Regarding this isotope, the 1.809 MeV γ-ray line emitted from the first excited state of ^{26}Mg, to which ^{26}Al decays, has been appointed as a tracer of the recent nucleosynthesis in our Galaxy and its presence has been found spread along Galactic plane. Observations support the idea that this isotope can been formed through nucleosynthesis in massive stars and core-collapse Supernovae [see 15, and references therein], but also Wolf-Rajet objects and AGB-stars [16], Novae [17] and X-ray burst [18] have been proposed as possible nucleosynthesis site. In any of those cases proton capture is allowed, and the ^{26}Al isotope can likely be produced via the reaction chain ^{24}Mg$(p,\gamma)^{25}$Al$(\beta^+)^{25}$Mg$(p,\gamma)^{26}$Al. A clear comprehension of the process is hampered by the presence of an isomeric state ^{26}Al^m($T_{1/2}$ = 6.34 sec), which can be directly fed by the ^{25}Al$(p,\gamma)^{26}$Si$(\beta^+)^{26}$Al^m chain, reducing the quantity of ^{26}Al in the interstellar medium. The ^{26}Si produced in this way can also be depleted by the ^{26}Si$(p,\gamma)^{27}$P reaction, that will therefore interfere with the production of ^{26}Al^m. The study of this reaction can than be useful to understand the ratio between ^{26}Al and ^{26}Al^m. Nonetheless, experiments involving ^{26}Si are challenging due to its short half-life

($T_{1/2}$ = 2.24 sec), and for this reason an indirect measurement of the ^{26}Si(p,γ)^{27}P has been performed using the ^{26}Mg(d,p)^{27}Mg one, applying the method described in Section 2 using mirror nuclei.
The experiment has been performed using the deuteron beam available at the U-120M isochronus cyclotron operated by the CANAM infrastructure of the Czech Academy of Sciences (Řež, Czech Republic) (E_{beam} = 19.2 MeV, I≈ 14 enA). Such a beam has been delivered on a ^{26}MgO (with ^{12}C as backing) target produced by the target laboratory of the Laboratori Nazionali del Sud - Istituto Nazionale di Fisica Nucleare (LNS–INFN).
The experimental apparatus was composed by five ΔE–E telescopes made using thin (250 μm) and thick (5000 μm) anular silicon detectors, covering the 7° – 60° range thanks to a rotating plate.
Once the Optical Model Parameters for the ^{26}Mg(d,p)^{27}Mg are extracted by means of the analysis of the angular distribution of the elastic scattering for the entrance channel and of the ^{26}Mg(d,p)^{27}Mg for the exit one, the ANC values for the ^{26}Mg + n → ^{27}Mg capture in the ground and first excited states have been found using Eq.6, in which the value of C^2_{pn} = 0.77 fm^{-1} [19] has been used. Once those quantities are known, the procedure for mirror nuclei has been applyied in order to extract the ANC and Γ_p for the ^{26}Si(p,γ)^{27}P

Table 1. Values found in [8] and compared with [20] and [21]

ref.	$(C^{Mg}_{g.s.})^2$ [fm^{-1}]	$(C^{Mg}_{1^{st}})^2$ [fm^{-1}]	$(C^{P}_{g.s.})^2$ [fm^{-1}]	$\Gamma^{P}_{1^{st}}$ [MeV]
[8]	28.26 ± 5.30	3.40 ± 0.32	1420 ± 255	(5.23 ± 1.05) × 10^{-9}
[20]	24.50 ± 4.90	1.1 ± 0.15	1058 ± 273	(4.04 ± 0.77) × 10^{-9}
[21]	44.00 ± 5.30	3.40 ± 0.32	1840 ± 240	(5.4 ± 0.1) × 10^{-9}

The results, shown in Table.1 show a substantial agreement between [8] and [20], if the different binding energies (0.807 MeV taken from [22] in [8] and 0.859 MeV in [20], respectively) are considered. Finally, once all the necessary quantities are extracted, the reaction rate for the direct and resonant capture in the ground and first excited state have been calculated using the formulas from [23], finding an enhancement by a factor 1.4 for the ground state and 2.2 for the first excited state contribution. To calculate the latter, the value of Γ_γ has been taken from [24].

3.2 The ^{6}Li(p,γ)^{7}Be and ^{3}He(α,γ)^{7}Be reactions

The ^{3}He(^{6}Li,d)^{7}Be reaction has been used to study both ^{3}He(α,γ)^{7}Be [9] and ^{6}Li(p,γ)^{7}Be [10] radiative capture reactions. This is possible because ^{6}Li and ^{3}He have a high probability to be found in the cluster configuration $\alpha \oplus d$ and $p \oplus d$, respectively.
The ^{3}He(α,γ)^{7}Be reaction is one of the most elusive ones in nuclear astrophysics. Despite the huge number of experimental and theoretical works performed to get information on the cross section during the last decades, its behaviour in the energy region of interest in the core of the Sun (between 10 keV and 30 keV at T = 15 MK) is still largely known only via theoretical extrapolations [25]. This reaction is important in H-burning in the Sun, given that it represents the first reaction of the so called 2nd and 3rd p-p chain branches. Also, a precise knowledge of the ^{3}He(α,γ)^{7}Be reaction would be useful in the context the long-standing lithium problem of Big Bang Nucleosynthesis (at energies around 100 keV) [26].
The ^{6}Li(p,γ)^{7}Be, on the other hand, represents one of the two main destruction channel for ^{6}Li along with ^{6}Li(p,α)^{3}He [27]. Given that the ^{6}Li is mainly produced via cosmic rays [28],

while ^{7}Li is formed in BBN and and in stellar scenarios [29], the isotopic ratio ^{6}Li/^{7}Li can be used to constraint the galactic enrichment process and lithium poroduction. Unfortunately, available experimental data are extracted at higher energies, leaving the behaviour of the cross section and reaction rate just to extrapolations in the region of astrophysical interest.
For the reason expressed above, both reactions have been studied in the same experimental campaign using the ^{3}He beam available at the 3.5 MV singletron accelerator of the Department of Physics and Astronomy (DFA) of the University of Catania (IT) an at the FN tandem accelerator at the John D. Fox Superconducting Accelerator Laboratory at Florida State University (Tallahassee, Florida - USA). Angular distributions have been studied at two beam energies (E_{beam} = 3 MeV and E_{beam} = 5 MeV), by means of movable silicon ΔE–E telescopes. The beam impinged on ^{6}LiF or on a pure ^{6}Li target evaporated on Formvar backing. At the backward angles in the center-of-mass reference frame, the differential cross section increases with the angle, hinting to the presence of a dominant one-step α-particle exchange mechanism – in this case the ^{3}He$(\alpha, \gamma)^7$Be reaction. In the forward hemisphere, on the other hand, this does not happen, and the one-step proton exchange process appears to be dominant. The ANC's for the $^3\mathrm{He} + \alpha \rightarrow ^7\mathrm{Be}$ and $^3\mathrm{He} + p \rightarrow ^7\mathrm{Be}$ have therefore been extracted separetely within the post form of the modified DWBA [12]. Channels coupling effects (CCE) have been taken into account considering the subsystem ^{3}He+^{6}Li, and coupling it with d+^{7}Be to study the proton transfer, and ^{7}Be+d to study the α-transfer. For the $^3\mathrm{He} + \alpha \rightarrow ^7\mathrm{Be}$, the calculation led to the extraction of $C^2 = 20.84 \pm 1.12$ fm^{-1} for the capture in the ground state and of $C^2 = 12.86 \pm 0.50$ fm^{-1} for the one in the 1st excited state (E_{ex} = 0.429 MeV), in agreement with [30]. The astrophysical S-factor at zero energy has been calculated as $S_{34}(0) = 0.534 \pm 0.025$ keV × b.
The ANC for the $^6\mathrm{Li} + p \rightarrow ^7\mathrm{Be}$ for the radiative proton capture in the ground and 1st excited state have been also extracted as $C^2 = 4.81 \pm 0.38$ fm^{-1} and $C^2 = 4.29 \pm 0.27$ fm^{-1}. The astrophysical S-factor at zero energy extracted using this values of the ANC is equal to $S(0) = 92 \pm 12$ eV × b, in agreement with [31], but with an uncertainty that is 1.6 times lower.

References

[1] A. M. Mukhamedzhanov and N. K. Timofeyuk, JETP Lett. **51**, 282 (1990)
[2] H. M. Xu et al., Phys. Rev. Lett. **73**, 2027 (1994)
[3] V Burjan et al., EPJ A **55**, 114 (2019)
[4] A. M. Mukhamedzhanov et. al., Phys. Rev. C. **84**, 024616 (2011)
[5] Y. P. Shen et al, Phys. Rev. C. **99**, 025805 (2019)
[6] N. K. Timofeyuk et al., Phys. Rev. Lett. **91**, 232501 (2003)
[7] L. Trache et al., Phys. Rev. C. **67**, 062801 (2003)
[8] G. D'Agata et al., Phys. Rev. C. **103**, 015806 (2021)
[9] G. Kiss et al., Phys. Rev. C. **104**, 015807 (2021)
[10] G. Kiss et al., Phys. Lett. B, **807**, 135606 (2020)
[11] A. M. Mukhamedzhanov & A. S. Kadyrov, Phys. Rev. C **82**, 051601 (2010)
[12] A.M. Mukhamedzhanov et al., Phys. Rev. C **56**, 1302 (1997)
[13] Typel, S. & Baur, G., Nucl. Phys. A **759**, 358-308 (2005)
[14] X.D. Liu et al., Phys. Rev. C **69**, 064313 (2001)
[15] L. Bouchet et al., Astrophys. J. **801**, 142 (2015).
[16] N. Prantzos and R. Diehl, Phys. Rep. **267**, 1 (1996)
[17] J. José et al., Astrophys. J. **560**, 897 (2001)
[18] O. Koike et al., A & A **342**, 464 (1999) (2004)

[19] R. V. Reid, Ann. Phys. **50**, 411 (1968)
[20] N. K. Timofeyuk et al., Phys. Rev. Lett. **91**, 232501 (2003)
[21] B. Guo et al., Phys. Rev. C **73**, 048801 (2006)
[22] L. J. Sun et al, Phys. Rev. C **99**, 064312 (2019)
[23] C. Iliadis, *Nuclear Physics of Stars* (Wiley-VCH Verlag, New York, 2007)
[24] L. J. Sun et al., Phys. Lett.B **802**, 135213 (2020)
[25] E.G. Adelberger et al., Rev. Mod. Phys. **83** 195 (2011)
[26] R.G. Pizzone et al., Astrophys. J. **786** 112 (2014)
[27] A. Gnech and L. E. Marcucci, Nucl. Phys. A **987**, 1 (2019)
[28] D. Clayton, *Handbook of Isotopes in the Cosmos* (Cambridge University Press, Cambridge, 2003).
[29] B. D. Fields et al., J. Cosmol. Astropart. Phys. **03**, 010 (2020)
[30] Q.I. Tursunmahatov and R. Yarmukhamedov, Phys. Rev. C **85**, 045807 (2012)
[31] D. Piatti et al., Phys. Rev. C **102**, 052802(R)(2020)

Nuclear astrophysics activities at the n_TOF facility at CERN

Cristian Massimi[1,2,*] *and* The n_TOF Collaboration[3]

[1]Department of Physics and Astronomy, University di Bologna, Bologna, Italy.
[2]Istituto Nazionale di Fisica Nucleare (INFN), Division of Bologna, Bologna, Italy
[3]www.cern.ch/ntof

Abstract. The n_TOF facility at CERN is operational since 2001, and provides neutron-induced cross section data of interest to several research fields, including nuclear astrophysics. The neutron time-of-flight (TOF) facility features three experimental areas located at different distances from the pulsed neutron source. Two beam lines at nominal distance of 185 and 19 m are especially equipped for TOF experiments. A third station at approximately 3 meters from the neutron source was conceived for irradiation and activation measurements. So far, neutron-induced cross sections for more than 100 isotopes have been measured.

1 Introduction

Radiative neutron capture, i.e. (n,γ), cross sections represent one of the most relevant nuclear inputs to models of stellar nucleosynthesis of the elements heavier than iron. For instance, the *s* process [1, 2] proceeds via a sequence of neutron captures and β-decays from a distribution of seed nuclei around iron, thus building up elements up to bismuth. In this scenario, β-decay rates are faster than neutron capture rates, the nuclear reactions proceed along the valley of stability on the chart of nuclei.

In addition to (n,γ) reactions, and to a minor extent, (n,p) and (n,α) reactions on a few light elements can play a relevant role when these reactions absorb a large number of neutrons, thus affecting the efficiency of the *s* process in synthesizing heavy elements. On the other hand, (n,p) and (n,α) are of some relevance in the modelling of the nucleosynthesis occurred during Big Bang or for the study of particular topics as the stellar production of the ^{26}Al gamma ray emitter observed in our galaxy.

The experimental observable of interest is the neutron-induced cross section averaged over the stellar neutron-energy distribution, typically referred to as Maxwellian averaged cross section (MACS). Experimentally, MACS are determined via two techniques: either time-of-flight (TOF) or activation. The TOF technique is based on the measurement of energy-dependent cross sections over a wide energy region, and subsequent calculation of the MCAS at different kT (k being the Boltzman constant and T the temperature). The second technique involves a sample first being irradiated with a neutron beam with a Maxwellian-like energy distribution, and subsequently the resulting product nucleus is counted. While from

*e-mail: massimi@bo.infn.it

TOF measurements MACS between $kT = 5$ and 100 keV can be estimated, the activation experiments are performed at a single temperature, typically around $kT = 30$ keV.

So far, the n_TOF collaboration has provided nuclear data using the TOF method for a large number of intriguing physics cases (see for instance ref. [3] and references therein). In the near future, the effort will be on activation measurements as well.

2 From laboratory experiments to stellar reaction rates

The astrophysical reaction rate depends on the number density of interacting particles times the reaction rate per particle pair $\langle \sigma v \rangle$. This latter term describes the probability of nuclear reactions between two particles, moving at relative velocity v. It is important to note that the interacting particles are in thermodynamic equilibrium in a stellar plasma, therefore their kinetic energy is linked to their thermal motion. Consequently as already mentioned, the relative velocity can be described by a Maxwell-Boltzmann distribution ϕ_{MB}:

$$\phi_{MB}(v)dv = \phi_{MB}(E)dE = \frac{2}{\sqrt{\pi}} \frac{1}{(kT)^{3/2}} \sqrt{E} e^{-\frac{E}{kT}} dE \qquad (1)$$

Figure 1 shows examples of Maxwell-Boltzmann energy distributions of neutrons in stars of different temperatures and the $^{25}Mg(n,\gamma)$ cross section for comparison.

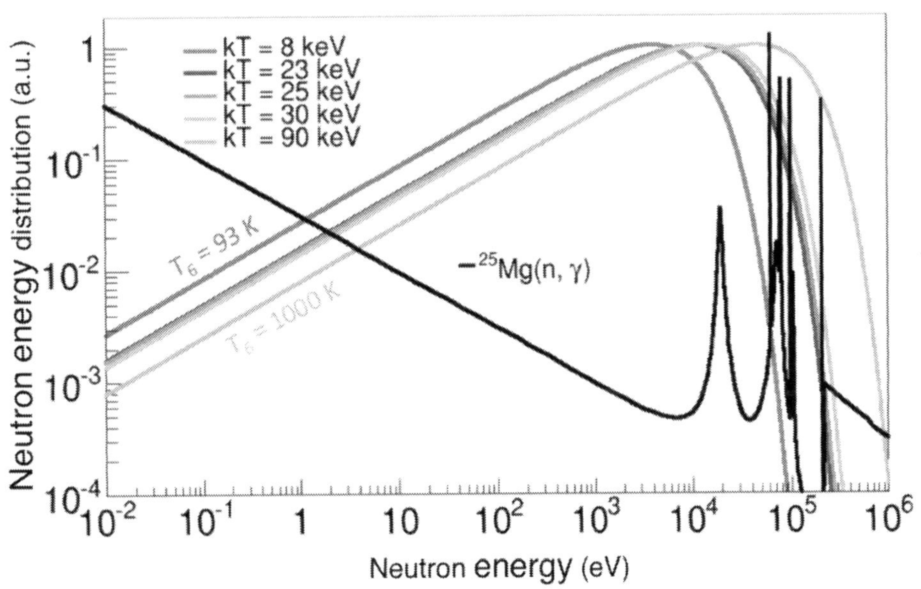

Figure 1. Example of energy distributions of neutrons in stellar interiors for different temperatures (colored curves), together with a (n,γ) cross section (black curve). For the sake of representation, y-scales are arbitrarily chosen.

From the same figure, it is evident that ϕ_{MB} presents a maximum located at different energy depending on the temperature. More in detail, the maximum occurs at the velocity $v_T = \sqrt{\frac{2kT}{\mu}}$, μ being the reduced mass of the system formed by the interacting particles. As neutron-induced reaction cross sections are measured as a function of energy E (here E

represents the centre-of-mass energy), it is customary to express the velocity distribution as energy distribution:

$$\langle \sigma v \rangle = \left(\frac{8}{\pi\mu}\right)^{1/2} \frac{1}{(kT)^{3/2}} \int_0^\infty E\sigma(E) e^{-\frac{E}{kT}dE}. \tag{2}$$

Therefore, the reaction rate can be also expressed in terms of the MACS:

$$\text{MACS} = \frac{\langle \sigma v \rangle}{v_T} = \frac{2}{\sqrt{\pi}(kT)^2} \int_0^\infty E\sigma(E) e^{-\frac{E}{kT}dE}. \tag{3}$$

The advantage of the TOF method with respect to activation is evident: numerical integration (Eq. 3) of energy-dependent cross section data makes is possible to derive MACSs at all relevant stellar temperatures ($5 < kT < 100$ keV). It is important to remark that the $\sigma(E)$ are measured over a large energy interval.

On the other hand, in case of rare and/or short-lived isotopes, it is not possible to prepare samples with sufficient mass (mg or higher) and/or enrichment for a TOF measurement. In these cases, an alternative to TOF measurements is the activation method, provided that neutron captures result in unstable isotopes. For instance, activation measurements on elements with half-life of minutes or less can be performed by cycling between irradiation and radiation measurement.

3 The n_TOF facility

After several facility upgrades, now n_TOF features two beam lines and corresponding experimental areas: EAR1 at 185 m and EAR2 at 19 m for TOF measurements, and an irradiation station referred to as NEAR for activation measurements.

The n_TOF facility is a white spallation source driven by the CERN proton synchrotron (PS). More in particular, neutrons are produced by 20 GeV/c protons from the PS, impinging onto a massive $80 \times 80 \times 60$ cm^3 Pb block [4]. The initially fast neutrons are moderated by a water layer of 5 cm, resulting in a wide neutron energy spectrum at both experimental areas EAR1 and EAR2 [5, 6]. Neutron energies span over 11 energy decades, from meV to GeV, and $\approx 5 \times 10^5$ and 10^7 neutrons per bunch reach EAR1 and EAR2, respectively. Finally, some 100 times higher neutron flux with respect to EAR2 is expected at NEAR. The high instantaneous neutron flux at relative large distances from the spallation target is the result of the combination of the PS features and the ones of the neutron-producing target.

3.1 Measurements at EAR1 and EAR2

The TOF technique is based on a measurement of the time needed by a neutron to travel a given distance L. This time t can be used to determine the neutron velocity: $v = L/t$, and consequently its kinetic energy E_n:

$$E_n = mc^2(\gamma - 1), \tag{4}$$

where γ represents the relativistic Lorentz factor $\gamma = (1 - v^2/c^2)^{-1/2}$, m is the mass of the neutron and c is the speed of light.

The PS provides a pulsed proton beam of approximately 10^{13} protons grouped in bursts of 7-ns FWHM. The primary proton beam produces a large amount of secondary particles, including γ-rays. These latter particles travelling along the beam pipe reach the experimental Areas after a fixed time $t_\gamma = L/c$, therefore providing a reference time to determine the "start" signal, i.e the moment when neutrons are produced in the spallation target. The "stop" signal is obtained from the time of detection of the neutron-induced reaction products. Consequently, the measured TOF is obtained from the time difference between stop and start signals.

In addition to the neutron flux, a discriminating feature of neutron facilities is the energy resolution:

$$\frac{\Delta E_n}{E_n} = \gamma(\gamma + 1)\sqrt{\left(\frac{\Delta L}{L}\right)^2 + \left(\frac{\Delta t}{t}\right)^2}, \quad (5)$$

where the resolution broadening is dominated by the neutron transport in the target-moderator assembly. Thanks to the long flight paths of EAR1 and EAR2, $\Delta E_n/E_n$ can be as small as 10^{-4}.

3.2 Measurements at NEAR

Being so close to the neutron-producing target, NEAR is characterized by a very high neutron flux, well suited for activation measurements of astrophysical interest. The neutron beam is transported from the spallation target to the NEAR station through a collimated pipe in the shielding wall, where samples are irradiated with $\approx 10^8$ neutrons per bunch.

This new area is complemented with a γ-ray spectroscopy laboratory equipped with an n-type HPGe detector of 55% relative efficiency, for the measurement of the activity resulting from irradiation of samples in the NEAR station.

Feasibility studies are ongoing to demonstrate the possibility of producing Maxwellian-like neutron spectra at different stellar temperatures by means of a neutron moderator/filter assembly, see for instance ref. [7].

In summary, after shaping the neutron spectrum to resemble a Maxwellian spectrum at a given temperature $\phi_{MB}(kT)$, a sample is irradiated for a certain period. The number of freshly produced nuclei are finally measured through the activity of the sample, which in first approximation is proportional to the MACS [8].

4 Nuclear Astrophysics at n_TOF

So far, the n_TOF collaboration has broadly studied different aspects of the *s*-process nucleosynthesis: (i) the main neutron sources in Red Giants stars (e.g. [9–12]); (ii) The seeds nuclei for the *s*-process (e.g. [13]); (iii) isotopes with a closed shell configuration (e.g. [14]); (iv) isotopes whose production is entirely ascribed to the *s* process (e.g. [15]); unstable isotopes where β decay competes with neutron capture (e.g. [16, 17]); (v) neutron poisons (e.g. [18, 19]); (vi) other reactions of interest, e.g. to Re/Os Cosmochronometer, to neutron standard [20].

Moreover, the n_TOF collaboration promoted a program of measurements aimed at studying neutron induced charged particle reactions of astrophysical interest, amongst them the measurement of ^{7}Be(n,α) and ^{7}Be(n,p) reactions relevant for Big Bang nucleosynthesis [21, 22], and neutron destruction of the cosmic gamma ray emitter ^{26}Al by (n,p) and (n,α) reactions [18, 19].

References

[1] E.M. Burbidge, G.R. Burbidge, W.A. Fowler, F. Hoyle, Reviews of Modern Physics **29**, 547 (1957)
[2] F. Käppeler, R. Gallino, S. Bisterzo, W. Aoki, Rev. Mod. Phys. **83**, 157 (2011)

[3] C. Massimi, S. Cristallo, C. Domingo-Pardo, C. Lederer-Woods, Universe **8** (2022)
[4] R. Esposito, M. Calviani, O. Aberle, M. Barbagallo, D. Cano-Ott, T. Coiffet, N. Colonna, C. Domingo-Pardo, F. Dragoni, R. Franqueira Ximenes et al., Physical Review Accelerators and Beams **24**, 093001 (2021), `2106.11242`
[5] M. Barbagallo, C. Guerrero, A. Tsinganis, D. Tarrío, S. Altstadt, S. Andriamonje, J. Andrzejewski, L. Audouin, V. Bécares, F. Bečvář et al., European Physical Journal A **49**, 156 (2013)
[6] M. Sabaté-Gilarte, M. Barbagallo, N. Colonna, F. Gunsing, P. Žugec, V. Vlachoudis, Y.H. Chen, A. Stamatopoulos, J. Lerendegui-Marco, M.A. Cortés-Giraldo et al., European Physical Journal A **53**, 210 (2017)
[7] G. Gervino, O. Aberle, A.P. Bernardes, N. Colonna, S. Cristallo, M. Diakaki, S. Fiore, A. Manna, C. Massimi, P. Mastinu et al., Universe **8** (2022)
[8] R. Reifarth, P. Erbacher, S. Fiebiger, K. Göbel, T. Heftrich, M. Heil, F. Käppeler, N. Klapper, D. Kurtulgil, C. Langer et al., European Physical Journal Plus **133**, 424 (2018), `1902.09838`
[9] S. Cristallo, M. La Cognata, C. Massimi, A. Best, S. Palmerini, O. Straniero, O. Trippella, M. Busso, G.F. Ciani, F. Mingrone et al., The Astrophysical Journal **859**, 105 (2018), `1804.10751`
[10] P. Adsley, U. Battino, A. Best, A. Caciolli, A. Guglielmetti, G. Imbriani, H. Jayatissa, M. La Cognata, L. Lamia, E. Masha et al., Physical Review C **103**, 015805 (2021), `2005.14482`
[11] C. Massimi, P. Koehler, S. Bisterzo, N. Colonna, R. Gallino, F. Gunsing, F. Käppeler, G. Lorusso, A. Mengoni, M. Pignatari et al., Physical Review C **85**, 044615 (2012)
[12] C. Massimi, S. Altstadt, J. Andrzejewski, L. Audouin, M. Barbagallo, V. Bécares, F. Bečvář, F. Belloni, E. Berthoumieux, J. Billowes et al., Physics Letters B **768**, 1 (2017)
[13] C. Lederer, C. Massimi, E. Berthoumieux, N. Colonna, R. Dressler, C. Guerrero, F. Gunsing, F. Käppeler, N. Kivel, M. Pignatari et al., Physical Review C **89**, 025810 (2014), `1403.4778`
[14] G. Tagliente, K. Fujii, P.M. Milazzo, C. Moreau, G. Aerts, U. Abbondanno, H. Álvarez, F. Alvarez-Velarde, S. Andriamonje, J. Andrzejewski et al. (n_TOF Collaboration), Phys. Rev. C **77**, 035802 (2008)
[15] A. Mazzone, S. Cristallo, O. Aberle, G. Alaerts, V. Alcayne, S. Amaducci, J. Andrzejewski, L. Audouin, V. Babiano-Suarez, M. Bacak et al., Physics Letters B **804**, 135405 (2020), `2002.02322`
[16] C. Lederer, C. Massimi, S. Altstadt, J. Andrzejewski, L. Audouin, M. Barbagallo, V. Bécares, F. Bečvář, F. Belloni, E. Berthoumieux et al., Physical Review Letters **110**, 022501 (2013), `1304.3310`
[17] C. Guerrero, J. Lerendegui-Marco, M. Paul, M. Tessler, S. Heinitz, C. Domingo-Pardo, S. Cristallo, R. Dressler, S. Halfon, N. Kivel et al., Physical Review Letters **125**, 142701 (2020)
[18] C. Lederer-Woods, P.J. Woods, T. Davinson, D. Kahl, S.J. Lonsdale, O. Aberle, S. Amaducci, J. Andrzejewski, L. Audouin, M. Bacak et al. (n_TOF Collaboration), Phys. Rev. C **104**, L022803 (2021)
[19] C. Lederer-Woods, P.J. Woods, T. Davinson, A. Estrade, J. Heyse, D. Kahl, S.J. Lonsdale, C. Paradela, P. Schillebeeckx, O. Aberle et al. (n_TOF Collaboration), Phys. Rev. C **104**, L032803 (2021)

[20] C. Massimi, C. Domingo-Pardo, G. Vannini, L. Audouin, C. Guerrero, U. Abbondanno, G. Aerts, H. Álvarez, F. Álvarez-Velarde, S. Andriamonje et al., Physical Review C **81**, 044616 (2010)

[21] M. Barbagallo, A. Musumarra, L. Cosentino, E. Maugeri, S. Heinitz, A. Mengoni, R. Dressler, D. Schumann, F. Käppeler, N. Colonna et al. (n_TOF Collaboration), Phys. Rev. Lett. **117**, 152701 (2016)

[22] L. Damone, M. Barbagallo, M. Mastromarco, A. Mengoni, L. Cosentino, E. Maugeri, S. Heinitz, D. Schumann, R. Dressler, F. Käppeler et al. (The n_TOF Collaboration [www.cern.ch/ntof]), Phys. Rev. Lett. **121**, 042701 (2018)

Radioactive Ion Beams: Production and Experiments at INFN-LNL

Marco Mazzocco[1,2*]

[1]Dipartimento di Fisica e Astronomia, Università di Padova, via F. Marzolo 8, 35131, Padova, Italy
[2]INFN-Sezione di Padova, via F. Marzolo 8, 35131, Padova, Italy

Abstract. In this contribution the main mechanisms and techniques for the production of Radioactive Ion Beams are reviewed. In particular, the in-flight facility EXOTIC at INFN-LNL will be described and the results of a recent measurement of interest for the cosmological ^{7}Li problem will be presented.

1 Introduction

In the chart of nuclei, the graphical representation of the combinations of protons and neutrons observed either in Nature or in experiments, radioactive nuclei outnumber stable nuclei by more than one order of magnitude [1]. As we move apart from the valley of β-stability, where stable and metastable nuclei are situated, we might encounter new phenomena, such as new decay modes, for instance proton-, neutron- and 2 proton-radioactivity, nuclei with exotic shapes or deformations, and also the well-established nuclear shell closures exhibit an evolution as we move towards the limits of existence on nuclei.

Radioactive nuclei are involved in many astrophysical scenarios, from the quiescent burning stages to the more dramatic supernova explosions. In our Sun, about 1% of the energy production is generated by the CNO cycle, which involves radioactive nuclei, and also the ppII and ppIII chains proceed through the intermediate production of unstable nuclei. In that respect, spectroscopic information on nuclei far from stability are of paramount relevance to calculate the cross sections for the reactions of interest and to predict the paths of the astrophysical processes in different stellar environments.

The study of the properties of the reactions induced by radioactive nuclei is complicated by the fact that, in most cases, these nuclei are short-lived, preventing the possibility of accumulating and safely handling macroscopic quantities useful for manufacturing a standard target or to be used as source material. Therefore, to perform these studies we need at first a (primary) nuclear reaction to occur for the production of the radioactive ion species of interest, which typically are not produced alone, thus we need also an efficient and fast selection technique to purify from the contaminant beams, in some cases even orders of magnitude more intense than the desired Radioactive Ion Beam (RIB). The rapidity of the entire process is another key requirement, since these ions are short-lived, and they will immediately start decaying after their production. Finally, dealing typically with rather small

* Corresponding author: marco.mazzocco@pd.infn.it

cross sections, all different stages for the RIB production must be optimized, from the choice of the most suitable projectile-target combination to the beam energy, from the power dissipation into the target to the design of instruments capable to operate in an environment potentially exposed to high radiation dose.

2 RIB Production Mechanisms

The production mechanisms for RIBs can be schematized into five categories: fragmentation, fission, spallation, fusion and two-body reactions in inverse kinematics.

2.1 Fragmentation

Fragmentation is caused by peripheral reactions leading to the removal of a certain number of nucleons. The production cross sections are nearly constant as a colliding energy of a few tens of MeV/u is exceeded. The advantage of increasing the beam energies, thus investing in the construction of a more powerful accelerator, consists in a smaller angular opening around the primary beam direction for the emission cone of the recoiling RIB under production. Fragmentation is mostly used for the production of proton-rich nuclei; however, it is often used as a powerful method to produce neutron-rich RIBs, especially in the heavier mass region.

2.2 Fission

Fission is a spontaneous decay mode for many transuranic isotopes. This process can also be induced by the energetic collision of fissile material with light nuclei. Due to the overwhelming abundance of neutrons with respect to protons in metastable fissile nuclei, this mechanism produces neutron-rich RIBs, especially in the mass region between A = 70 and A = 160. Due to the systematics of fission reaction, the emission cones for the fission fragments are significantly larger than for fragmentation reaction products. Emission cones with wider angular opening require the design of particle spectrometers with larger angular acceptance to guarantee a better transmission and, in turn, a higher RIB intensity.

2.3 Spallation

Spallation is the interaction of a particle, mostly a single nucleon, at energies between 100 MeV and a few GeV with a target. After the interaction, many nucleons are ablated and the residual nuclei can be very different from the initial target nucleons. This mechanism has rather high cross sections for the production of neutron-deficient RIBs.

2.4 Fusion

Fusion is a process where, after the interaction, projectile and target form a compound nucleus at moderately high excitation energy. The system de-excites by emitting protons, neutrons, alpha particles, and gamma rays giving origin to a plethora of evaporation residues. These nuclei recoil in forward direction with a mean velocity approximately equal to the compound nucleus velocity, which is a fraction significantly smaller than the projectile velocity. Recoil separators exploit this feature to select the residues, however the separation efficiency is strongly limited, since this mechanism occurs essentially in the low energy regime (a few MeV/u) and consequently the residues are produced with broad angle, velocity and charge state distributions. So far, fusion has been employed for spectroscopy studies, for the synthesis of superheavy elements, but not for the production of RIBs.

2.5 Two-body reactions in inverse kinematics

We group into this category inverse (p,n), (d,n) and (^{3}He,n) reactions, preferably with negative Q-value and forward-peaked differential cross sections. Linear momentum conservation ensures that the RIBs are produced with narrow openings (< 10°) around the primary beam direction. Hydrogen and helium are the preferred target material for this mechanism, but there are no helium compounds and hydrogen-rich compounds cannot typically withstand the thermal stress induced by an intense primary beam. Therefore, to circumvent these problems, the use of gas targets with metal windows is routinely undertaken. This production mechanism works particularly nicely for light RIBs (up to A = 30-40) a few mass units away either on the n-rich or p-rich sides of the valley of stability.

3 RIB Production Techniques

Unless the radioactive species of interest is long-lived enough and has the suitable chemical properties to be produced in batch-mode and employed as ion source or target material, there exist two complementary techniques for the RIB production and separation: the in-flight and the Isotope Separation On-Line (ISOL) methods.

3.1 In-Flight

In the in-flight method [2], a heavy-ion beam with energy ranging from a few MeV/u to a few GeV/u impinges on a thin production target. For projectile fragmentation and fission, the preferred target material is beryllium, a monoisotopic element with low atomic number and high melting point. The inverse kinematics and the high bombarding energy guarantee a strong forward focusing at small angles around the primary beam direction of the reaction products.

The RIB selection exploits proper combinations of electromagnetic fields and atomic interaction, occurs in a fast time scale (the time needed to travel through the entire separator, usually less than 1 μs) and is independent from the RI chemical properties. Moreover, the secondary beams retain a large fraction of the initial projectile velocity, thus in-flight facilities usually do not require the construction of a post-accelerator. The use of a thin target is recommended to minimize the energy loss and the angular straggling through the target thickness, which might compromise the performances of the fragment separator in terms of transmission, selection quality and purification of the RIB under production. Possible drawbacks of this technique are associated with the generally poor beam quality, i.e. large longitudinal and transverse emittance and beam-spot on the secondary target. Nevertheless, this method is very fast and universal, RIBs are directly available at high energy and is particularly adapted for the production and identification of new isotopes.

3.2 ISOL

The ISOL [3] technique provides very good quality beams with small emittance and energy spread and with high purity. These positive aspects come at the price of the complexity of the entire facility. In fact, the primary beam impinges on a thick production target. The produced radioactive species diffuse into the target material, kept at high temperature to increase the ion mobility, then effuse into the target container and, after a series of collision with the walls, are finally transferred towards the source to be ionized. All these processes (diffusion, effusion, sticking time on the container walls, extraction, transfer, ionization) are strongly chemistry dependent and require a certain amount of time. Compared to the in-flight method, ISOL is considered to be significantly slower and long development times are needed for each different ion species.

After the ionization, which might already be a rather selective process, especially when laser frequencies are employed, the radioactive ions are ready for a isotopic/isobaric separation, whose resolution depends on the optical design of the magnetic spectrometer. At this stage, the RIB is still at low energy and with a small charge state. To match the beam energy range of interest for "standard" Nuclear Physics experiments, a post-acceleration stage must be implemented in an ISOL facility. To increase the performance of the post-accelerator, the RIB charge state can be raised by means of a charge breeder. At the end of the post-acceleration, the RIBs have practically the same optical properties of stable beams and are particularly adapted for detailed spectroscopic studies.

4 The Facility EXOTIC

EXOTIC [4] is an in-flight facility installed at the Laboratori Nazionali di Legnaro (LNL) of the Istituto Nazionale di Fisica Nucleare (INFN), near Padova (Italy). Light weakly-bound RIBs are produced by means of two-body inverse kinematic reactions induced by heavy-ion beams, accelerated by a 15 MV XTU-Tandem, impinging on a gas target. The commissioning of the facility was performed in 2004 and a substantial upgrade process was subsequently held in 2012 [5]. So far, 8 light RIBs have been delivered in the energy range of a few MeV/u: ^{8}Li, ^{7}Be, ^{8}B, ^{10,11}C, ^{15}O, ^{17}F and ^{18}Ne.

The ion optical design of EXOTIC includes a first quadrupole triplet, a dipole magnet, a Wien filter and a second quadrupole triplet. The double selection in magnetic rigidity and velocity acted by the dipole magnet and the Wien filter, respectively, helped achieving secondary beam purities as good as 99%. Mostly reaction dynamics studies at Coulomb barrier energies and experiments on clustering have been performed. In March 2021 the two reaction chambers at the final focal of EXOTIC were dismounted to permit the installation of the Advanced GAmma-ray Tracking Array (AGATA [6]) and platform with the relative services and we submitted a Technical Design Report and a Letter-of-Intent for investigating the possibility to couple the facilities EXOTIC and AGATA and exploit the unique features guaranteed by the light RIBs delivered and the state-of-the-art performances of AGATA.

5 Recent Experiment

Quite recently, a first experiment [7] of astrophysical interest was performed to study the ^{7}Li cosmological problem. In fact, the ^{7}Li abundance observed in metal-poor stars is a factor of 3 smaller than standard Big-Bang Nucleosynthesis (BBN) theoretical predictions. In the BBN reaction network, ^{7}Li is essentially produced by the electron capture decay of ^{7}Be. In that respect, the investigation of the ^{7}Be destruction mechanisms in the BBN energy range (20-70 keV) might provide a Nuclear Physics solution of the long-standing ^{7}Li cosmological problem [8]. The second most dominant ^{7}Li destruction process, after the ^{7}Be(n,p)^{7}Li, is the reaction ^{7}Be(n,α)^{4}He. The study of this reaction is complicated by the fact that it involves both a radioactive projectile and a radioactive target. We therefore employed the Trojan Horse (indirect) Method (THM) [9] by making use of the surrogate three-body reaction d(^{7}Be,α^{4}He)p.

The ^{7}Be RIB was delivered by the facility EXOTIC at 20.4 MeV with an intensity of 5-8 x 10^{5} pps and impinged on a 400-μg/cm^{2} CD_2 target. The two α-particles were detected in coincidence by means of four telescopes of the EXPADES detector array [10], placed at suitable angles to fulfil the requirements prescribed by the THM for the observation of a quasi-free process, where the (undetected) proton acted as a spectator in the reaction. The trend of cross section extracted for the ^{7}Be(n,α)^{4}He with the THM is in agreement with the recent direct measurement (of the reverse reaction) performed by Kawabata et al. [11] and with the work of Hou and collaborators [12].

In conclusion, the discrepancy between the predicted value for the ^{7}Li abundance and the observed value for low-metallicity stars still remains large, suggesting once again the need of alternative solutions for the cosmological lithium problem. However, we should underline that this study represents one of the first applications of the THM to RIB induced reactions, following the main stream started with the investigation of the reaction ^{18}F(p,α)[13-15]. Nevertheless, ^{7}Be(n,α)^{4}He represents the first attempt to measure via the THM a reaction involving a neutron and a RI, similarly to the study recently carried out for the reaction ^{7}Be(n,p)^{7}Li [16], also of paramount importance for the BBN. These works open the way to the possible future investigation of other neutron induced reactions of astrophysical interest with short-lived radioactive nuclei exploiting the capabilities of the THM.

References

1. https://people.nscl.msu.edu/~thoennes/isotopes/index.html
2. D.J. Morrissey and B.M. Sherill, Lect. Notes Phys. **651**, 113 (2004)
3. P. Van Duppen, Lect. Notes Phys. **700**, 37 (2006)
4. F. Farinon et al., Nucl. Instrum. and Methods B **266**, 4097 (2008)
5. M. Mazzocco et al., Nucl. Instrum. and Methods B **317**, 223 (2013)
6. https://www.agata.org
7. L. Lamia et al., Ap. J. **879**, 23 (2019)
8. C. Broggini et al., Journal of Cosmology and Astroparticle Physics **6**, 030 (2012)
9. R.E. Tribble et al., Reports on Progress in Physics **77**, 106901 (2014)
10. D. Pierroutsakou et al., Nucl. Instrum. and Methods A **834**, 46 (2016)
11. T. Kawabata et al., Phys. Rev. Lett. **118**, 052701 (2017)
12. S.Q. Hou et al., Phys. Rev. C **91**, 055802 (2015)
13. S. Cherubini et al., Phys. Rev. C **92**, 015805 (2015)
14. R.G. Pizzone et al., Eur. Phys. J. A **52**, 24 (2016)
15. M. La Cognata et al., Ap. J. **846**, 65 (2017)
16. S. Hayakawa et al., Ap. J. **915**, L13 (2021)

Basic Stellar Physics

S. Palmerini[1,2,3]

[1] Dipartimento di Fisica e Geologia, Universitá degli Studi di Perugia, 06123 Perugia, Italy
[2] Istituto Nazionale di Fisica Nucleare, Sezione di Perugia, 06123 Perugia, Italy
[3] Istituto Nazionale di Astrofisica, Osservatorio Astronomico di Roma, Monte Porzio Catone, Italy

Received: date / Revised version: date

Abstract. Polytropes, Virial theorem, evolutionary time scales, degrees of freedom, radiative transport, molecular weight, degeneracy, Jeans mass, and Eddington luminosity are basic ingredients to describe the physics of stars. In the present paper they will be presented in details as long with their role in the stellar evolution.

PACS. PACS-key stellar evoluion – PACS-key basic physics

1 Introduction

This is not a scientific paper dealing with research results, it is instead thought as a document devoted to students of the European Summer school on Experimental Nuclear Astrophysics who might keep it in their virtual portfolio and consult it, when needed during their present and future studies. Stars evolve by going through successive stages of equilibrium (or non-equilibrium) between the contraction driven by gravity and expansion due to the pressure of gas and radiation fed by thermonuclear burning. Several papers and books describe in details the stellar evolution and nucleosynthesis processes, but the basic ingredients describing those phenomena are few: polytropic relations, Virial theorem, time scales, Jeans mass, degrees of freedom, radiative transport, molecular weight, Eddington limit and degeneracy. In the following paragraphs we will describe how each of these elements of basic physics plays a crucial role in the stellar formation as well as in other stages of the evolution.

2 Polytropes

Let us assume for simplicity that stars are made of perfect gases, therefore from the equation of state the gas pressure P can be cexpressed as

$$P = nRT/V \tag{1}$$

where $R = kN_A = 8.3144621 Jmol^{-1}K^{-1}$ is the ideal gas constant, n the number of moles, $k = 1.3806 \cdot 10^{-23} J/K$ the Boltzmann's constant, and $N_A = 6.022 \cdot 10^{23}$ the Avogadro's number. By writing the volume V as a function of the gas density in equation 1 and using the definition of the ideal gas constant, R, the pressure P turns out to be:

$$P = \frac{k}{\mu m_H} \rho T \tag{2}$$

where μ is the molecular weight and m_H is the atomic mass unit.

We can here immediately recognize a particular case of a polytropic transformation [1]. From equation 2 it is clear that the pressure is proportional to the density ρ whenever the temperature T is constant. Therefore, one can write that $P \propto \rho^\delta$, being $\delta = 1$ in case of isothermal processes, where $dT = 0$ and c (the specific heat) is equal to 0. Another special case is the one of adiabatic processes for which $\delta = \gamma$, where γ is the adiabatic exponent namely the ratio between the isobaric heat capacity and the isochoric one.

According with Chandrasekhar 1958 [2], a polytrope describes a thermodynamic transformation of a perfect gas, where the energy exchange is proportional to the variation in the temperature: $\delta Q = cdT$, being T the temperature in Kelvin degrees. If the energy is preserved ($\delta Q = 0$) the equation of the adiabatic transformations holds:

$$P \propto A = \pi r^2 V^{-\gamma} \rightarrow P \propto \rho^{\gamma} \tag{3}$$

where the adiabatic exponent γ t is equal to 5/3 for a perfect gas with 3 degrees of freedom. γ is defined by relation 3, which according with equation 1 can also be written as

$$T \propto \rho^{\gamma-1} \tag{4}$$

By differentiating equations 3 and 4 one obtains

$$\frac{dT}{T} = \gamma - 1\frac{d\rho}{\rho} \tag{5}$$

$$\frac{dP}{P} = \gamma\frac{d\rho}{\rho} \tag{6}$$

Suppose a nebula contracts till equilibrium is reached. Then the pressure gradient equals the gravitational potential:

$$\frac{dP}{dr} = -\rho g = -G\rho\frac{M}{r^2} \tag{7}$$

and for a system with a spherical symmetry

$$dP = -G\frac{4\pi}{3r^2}\rho^2 r^3 dr = -G\rho^2 r\frac{4\pi dM}{12\pi\rho r^2} \rightarrow P \propto \rho^{\frac{4}{3}}. \tag{8}$$

Any stable thermal pressure behaves as a polytrope of index $\delta \geq \frac{4}{3}$, while δ is smaller than $\frac{4}{3}$ in collapsing systems. In statistical mechanics, q degrees of freedom imply that $\gamma = 1 + \frac{2}{q}$. Then if $q = 3$ γ turns out to be $\frac{5}{3}$ while $q = 6$ leads to $\gamma = \frac{4}{3}$. Whenever q is larger than 6 the structure is not stable. Furthermore, $\delta = \frac{n+1}{n}$, being n the polytropic index and so $n = \frac{3}{2}$ implies $\delta = \frac{5}{3}$ and $n = 3$ $\delta = \frac{4}{3}$.

The above discussion holds for a perfect gas, whose pressure is only the pressure of the gas itself. Instead, in a stellar *plasma* the gas pressure is just a fraction β of the total one and usually the radiation pressure is the remaining part $(1 - \beta)$. For an adiabatic transformation of a plasma, admixture of gas and radiation, relations similar to the one valid for a prefect gas hold, but the exponent is no longer equal to a single γ value and three adiabatic exponents Γ_i have to be defined for transformations in presence of radiation:

$$\frac{dP}{P} = \Gamma_1\frac{d\rho}{\rho} \tag{9}$$

$$\frac{dP}{P} = \frac{\Gamma_2}{\Gamma_2 - 1}\frac{dT}{T} \tag{10}$$

$$\frac{dT}{T} = (\Gamma_3 - 1)\frac{d\rho}{\rho} \tag{11}$$

Hence

$$\frac{dlnP}{ln\rho} = \Gamma_1 \tag{12}$$

$$\frac{dlnP}{dlnT} = \frac{\Gamma_2}{\Gamma_2 - 1} \tag{13}$$

$$\frac{dlnT}{dln\rho} = (\Gamma_3 - 1) \tag{14}$$

It can also be shown that when $\beta < 1$, also the molecular weight μ is variable and one can define its gradient:

$$\nabla_\mu = \frac{dlnT}{dln\mu} \tag{15}$$

Relations from 9 to 15 are useful in stellar modeling to define the borders between regions where radiation or convection dominate.

3 The Virial theorem

Let's consider an interstellar cloud made of many particles of masses m_i, in a reference frame where r_i is the distance vector of each particle m_i from the origin. Each particle is subject only to internal forces, namely gravity, and such a system is called a "virial" system (Clausius 1870 [3]). Its energy is defined as:

$$\sum_{i=1} m_i \frac{d^2 r_i}{dt^2} = -G \sum_{i=1} \frac{M m_i}{r_i^2} \tag{16}$$

By scalary multiplying by the vector r_i one obtains:

$$\sum_{i=1} m_i r_i \ddot{r_i} - G \sum_{i=1} \frac{M m_i}{r_i} \equiv \Omega \tag{17}$$

By definition the moment of inertia and the kinetic energy of each particle are $I_i = m_i r_i^2$ and $E_k = \frac{1}{2} m_i \dot{r_i}^2$, respectively. The whole cloud moment of inertia is $I = \sum_{i=1} m_i r_i^2$, while the total kinetic energy is equal to the thermal energy $E_k = U \equiv \frac{3}{2} nkT$. By deriving I twice the following relation is easily found:

$$\sum_{i=1} \frac{d^2(m_i r_i^2)}{dt^2} = 2m\dot{r}^2 + 2mr\ddot{r} \tag{18}$$

Hence

$$\frac{1}{2}\frac{d^2 I}{dt^2} = 2E_k + \Omega \tag{19}$$

where $m = \sum_{i=1} m_i$ and when the cloud reaches equilibrium:

$$E = \frac{1}{2}\Omega + \Omega = U - 2U \rightarrow E = -U \tag{20}$$

Any increasing E results in a decrease in U (and T) and self-gravitating clouds have a negative specific heat.

4 Evolutionary time scales

Equation 17 can be written for the unit volume so that $m = \rho$. Then the cloud starts heating due to contraction (equation 20), and a pressure gradient starts to appear (see [1]), so that the force balance becomes:

$$\rho \frac{\partial^2 r}{\partial t^2} + \frac{\partial P}{\partial t} = -\frac{G(M_r \rho)}{r^2} \tag{21}$$

At the very beginning the pressure gradient is minimal, and one can approximately assume: $\frac{dP}{dr} \simeq 0$, then the cloud starts to contract. Let us make the hypothesis, for the sake of simplicity, that the velocity (of each particle) is proportional to the distance from the centre ("homologous" collapse). Assuming an average density ρ_0 for the nebula its mass will be $M_r \equiv (M_r)_0 = \frac{4}{3}\pi\rho_0 r_0^3$: hence one can easily compute the the free fall time scale, namely the time required for the unimpeded collapse driven by gravity:

$$\tau_{ff} = \left(\frac{3\pi}{32 G \rho_0}\right)^{\frac{1}{2}} \rightarrow \tau_{ff} \simeq \frac{1}{\sqrt{G\rho}} \tag{22}$$

By assuming $M_r = M_\odot$ and $\rho = \rho_{\odot,average}$, the free fall time is about 1800s, namely the Sun would evolve in only half an hour. Therefore it is evident that another source of energy is needed to feed stars and ensure a time scale of billions of years for their evolution. This was the starting point for the thermodynamical analysis by Kelvin and Helmoltz. According with them, the total energy of a star E is constant, only half of the gravitational energy goes into keeping the gas warm and, if we have a prefect gas, with only thermal degrees of freedom, half gravitational energy can be radiated [4].

Now let's consider the gravitational potential of a gaseous sphere

$$\Omega = -\int_0^{M_*} G\frac{M dM}{r_M} = -\frac{3}{5} G \frac{M_*^2}{R} \tag{23}$$

being R and $M_\star$ the stellar radius and mass, respectively. The energy that can be radiated is half of the value resulting from equation 23. If it is emitted at the present solar rate ($3.8 \times 10^{33} erg/sec$), then the Kelvin-Helmoltz time scale is:

$$\tau_{KH} = \frac{3GM_\odot^2}{10R_\odot L_\odot} \simeq 2 \cdot 10^7 yr \tag{24}$$

In 1920 Rutherford and Holmes published the results of several years of measurements on ancient rocks [5]. From the abundance of actinides they derived for the Earth an age larger than 1 Gyr. That was the proof that gravity cannot be the main source of energy for the Sun: it would be much younger than the Earth itself.

Firstly, sir Eddington [6] suggested that a sufficiently long lifetime would be guaranteed to the Sun by the fusion of 4 protons into a He nucleus. The binding energy of ^{4}He is 26.71 MeV (the 0.7% of the mass of 4 protons) then the energy delivered by the Sun by burning about 10% of its H mass can be calculated starting from the Einstein's equation:

$$E = mc^2 \to \Delta E = 0.1 \cdot 1.989 \cdot 10^{33} \cdot 0.007 \cdot (2.99 \cdot 10^{10})^2 \simeq 1.3 \cdot 10^{51} erg \tag{25}$$

Being the present solar luminosity (namely the energy radiated per unit of time) $L_\odot = 3.8 \cdot 10^{33} erg/s$ the time scale for solar evolution powered by H-burning is

$$\tau_{nucl} = \frac{1.3 \cdot 10^{51}}{3.8 \cdot 10^{33}} = 3.4 \cdot 10^{17} s \to \tau_{nucl} \geq 10^{10} yr \tag{26}$$

Indeed, $4.603 \cdot 10^9 yr$ is the estimated age of the Sun, which is supposed to spend in total about $0.8 \cdot 10^{10} yr$ on the Main Sequence, burning H in its core. In 1928 Gamow [7] showed, by solving Schroedinger's equation in a simple exercise, that he Tunnel Effect allows the Sun to fuse protons and synthesize helium. In 1938 Bethe and Critchfield [8] published a detailed list of the nuclear reactions, which operate the transmutation of H into He in stellar environments, the so called pp-chain.

5 The Jeans' mass

Beside a too short stellar lifetime, another problem related to the Virial theorem is that it would allow the formation only of very massive objects. Applying equation 16 to the interstellar medium one obtains that the Jeans mass, namely the minimum mass needed to form a star (in equilibrium),is:

$$M_J = \left(\frac{5k}{G\mu m_H}\right)^{\frac{3}{2}} \left(\frac{3}{4\pi\mu m_H}\right)^{\frac{1}{2}} \frac{T^{\frac{3}{2}}}{n^{\frac{1}{2}}} \simeq 2 \cdot 10^{35} \frac{T^{\frac{3}{2}}}{n^{\frac{1}{2}}} \tag{27}$$

By assuming $T = 10K$ and a density n of 10 particle per cm^3, M_J is about $2 \cdot 10^{36} g \simeq 1000 M_\odot$, which is the order of magnitude of the masses of open clusters and not of a single star. According to equation 27 massive objects of 20-30$M_\odot$ may form in the compact cores of interstellar medium clouds, where $T \sim 20 - 30K$ and $n \geq 10^8 cm^{-3}$. However, the smaller is the mass the larger is the number of stars with that mass in a galaxy, then how could the majority of the stars form? They form by fragmentation triggered by forces external to the nebula. The cloud cores actually form by a slow accretion induced by extra pressures (galactic dynamics, magnetic fields, etc...). In these case equation 20 is:

$$2U + \Omega = 3 \int PdV \tag{28}$$

where $3 \int PdV$ is the external work spent in fragmentation. According to the Virial theorem, when the cloud is in thermal equilibrium half of its gravitational energy can br used to break solids and molecules (non thermal degrees of freedom) and when the density ρ increases, M_J rapidly decreases allowing the formation of typical stellar masses. Every perturbation of ρ induces fragmentation and smaller mass clusters form. If these fragments reach the virial equilibrium they can be described by polytropes with $\delta \geq 4/3$. For them:

$$T \propto \rho^{\delta-1} \to M_J \propto \rho^{\frac{3\delta-4}{2}} \to M_J \propto \rho^{\frac{1}{2}} \tag{29}$$

The denser the fragments the more massive the formed stars are.

6 Radiative transport

When energy is transported by the interaction of photons with matter (photons being absorbed and re-emitted by each stellar layer, with a very short mean free path) the fraction of the flux $d\Phi_\nu$ across a layer dr will be proportional to the flux Φ_ν itself.

$$d\Phi_\nu = -\Phi_\nu k_\nu \rho dr \rightarrow \frac{d\Phi_\nu}{dr} = -k_\nu \rho \Phi_\nu \tag{30}$$

where the constant k_ν deals with interactions between radiation and matter per unit volume. Since $dP_r = \frac{d\Phi_\nu}{c}$ and $\Phi_\nu = \frac{L}{4\pi r^2}$:

$$\frac{dP_r}{dr} = -\frac{k_\nu \rho L(r)}{4\pi c r^2} \tag{31}$$

7 The molecular weight

In a unit volume (whose mass is ρ) the number of particles is equal to the mass of the gas divided by the average mass of each particle:

$$n = \frac{\rho\mu}{m_H} \tag{32}$$

Stellar composition is often simply defined by the parameters X, Y and Z, being X is the mass fraction of hydrogen, Y the mass fraction of helium, and Z the metallicity, namely the mass fraction of the heavier elements ($A > 4$). Thus $X + Y + Z \sim 1$, but in a stellar plasma the particles include ions and free electrons and every fully ionized hydrogen atom releases one electron, hence the particles associated to H are:

$$n_H = 2X\frac{\rho}{m_H} \tag{33}$$

Every He ion has mass 4 m_H and delivers 2 electrons, whose masses can be neglected being much smaller than the nuclear one, but the number of particles is 3, then

$$n_{He} = \frac{3}{4}Y\frac{\rho}{m_H} \tag{34}$$

Finally for heavier elements (the so-called metals), the most abundant ones are those made of α particles (^{12}C, ^{16}O, ^{20}Ne...), for which the charge is half the atomic mass ($A/2$). Hence for any nucleus of mass A_i:

$$n_{A>4} = \frac{A}{2} \cdot \frac{1}{A}Z\frac{\rho}{m_H} = \frac{Z}{2}\frac{\rho}{m_H} \tag{35}$$

The average molecular weight μ for a stellar plasma can then be computed combining equation 33, 34, and 35:

$$\mu = \frac{\rho}{m_h n} = \left[2X + \frac{3}{2}Y + \frac{1}{2}Z\right]^{-1} \tag{36}$$

Values of the average molecular weight vary depending on the stellar environment under analysis. For a pure hydrogen gas ($X = 1$) $\mu = 0.5$, while for a mixture of pure heavy elements $\mu = 2$, as for example in the innermost cores of massive stars before the core-collapse of a supernova. A ionized gas with the solar composition, where the total abundance of elements heavier than 4 is the solar metallicity Z=0.015 (X = 0.71 and Y = 0.275) $\mu \simeq 0.612$. For electrons we have one particle per H nucleus, one-half per He nucleus and many electrons per ions of heavy elements. Hence:

$$\mu_e \simeq \left[X + \frac{Y}{2} + \frac{1}{2}(1 - X - Y)\right]^{-1} \tag{37}$$

8 Degeneracy

Degeneracy prevents stars or protostars to evolve. Indeed whenever the plasma is degenerate the on-set of the nuclear burning typical of the next evolutionary stage is hampered. It is so for clouds with mass smaller than $0.08M_{\odot}$, which will never reach the physical condition needed to burn H in their core, or for red giants whose He core is smaller than $1.4M_{\odot}$ (which is allowed to burn just by a late instability called He-flash), or for any other evolved object that is not massive enough to promote a further nuclear burning in its core (e.g. C-burning after the He-burning, O-burning after C-burning, etc...).

In a simplified way one can state that the inner layers of a star (or a cloud) become degenerate when the temperature does not grow enough to allow the fusion between charged particles, being its mass too low. As already explained, when gravity dominates the density increases, because only half gravitational potential energy is lost through radiation and no nuclear energy is provided to compensate the rest. This is so as during the pre-main sequence or at the end of each nuclear burning phase along stellare evolution. Since the stellar plasma is not made of ideal gases, under those conditions particles become in contact with one another, each one occupying a volume l^3 where l is the mean free path $l \sim \sqrt[3]{n^{-1}} = \sqrt[3]{m/\rho}$. The length l can be easily estimated from the De Broglie wave length $\lambda_{DB} = h/mv$, here h is the Plank constant and v the particle thermal velocity in a Maxwell distribution, $v \simeq \sqrt{3kT/m}$. By some simple algebra the critical density above which the system will be degenerate turns out to be:

$$\rho_{deg} \propto \sqrt{\frac{T^3}{m^5}} \tag{38}$$

For T values typical of stars that have just left the Main Sequence (a few $10^7 K$) and for particles of mass equal to the mass of electrons, the critical density is about $10^6 g/cm^3$. Stars, which encounter electron degeneracy before settling on the Main Sequence, evolve at very high central density and radiate their gravitational energy with no H-burning. Such objects are called brown dwarfs (and giant planets as Jupiter are at the lowest). For particles more massive than electrons ρ_{deg} is larger: electrons become a degenerate gas at a density much lower than the ones at which protons or neutrons do , typically by a factor $2000^{5/2}$, according to equation 38.

9 The Eddington Limit

In the same way the degeneracy fixes the lower limit of the stellar masses the luminosity fixes its upper limit, called the Eddington limit [9]. Indeed, radiation pressure on electrons and ions at a stellar surface is proportional to its luminosity and it drives stellar winds. When radiation pressure exceeds the gravitational one the star approaches the Eddington luminosity L where radiation pressure would destroy it.

$$L = \frac{4\pi G c m_p M}{\sigma_e} = 1.26 \times 10^{38} \frac{M}{M_{\odot}} erg \cdot s^{-1} \tag{39}$$

Equation 39 gives the maximum luminosity that an object of mass M emitting its radiation isotropically can have, above this limit the star would be blown away by its own radiation pressure. Therefore, by inverting the formula, one obtains the mass upper limit for a star with a given luminosity L:

$$M_E = 8 \times 10^5 \frac{L}{10^{44} erg \cdot s^{-1}} M_{\odot} \tag{40}$$

which corresponds to a limiting mass slightly lower than 200 $M_{\odot}$.

References

1. S. Chandrasekhar, An Introduction to the Study of Stellar Structure, Dover Books on Astronomy (2010)
2. S. Chandrasekhar, Astrophys. J **128**, (1958) 114.
3. R. Clausius, Annalen der Physik **217** 9 (1870) 124
4. R. Kippenhahn, A. Weigert, A. Weiss, Achim, Stellar Structure and Evolution, Springer Berlin Heidelberg (2013)
5. D. Palmer, Nature **409** (2001) 20.
6. A. S. Eddington, Nature **132** 3338 (1933) 639.
7. G. Gamow, Nature **122** 3082 (1928) 805.
8. H. A. Bethe, C. L. Critchfield, Physical Review **54** 10, (1938) 862.
9. A. S. Eddington, Nature **113** 2848 (1924) 786.

Investigation of a light Dark Boson existence : The New JEDI project

Beyhan Bastin[1,*,**,***], *Jürgen* Kiener[2,**,***], *Isabelle* Deloncle[2,**], *Alain* Coc[2,**], *Maxim* Pospelov[3,**], *Jaromir* Mrazek[4,**], *Livio* Lamia[5,6,**], *Dieter* Ackermann[1], *Philip* Adsley[7,8,***], *Charles-Olivier* Bacri[2], *Jérôme* Bourçois[2], *Vaclav* Burjan[4], *Anastasia* Cassisa[4], *Giuseppe* D'agata[4,5,6], *Gilles* De France[1], *Alessia* Di Pietrio[6], *Yasmine* Demane[9], *François* De Oliveira[1], *Lindsay* Donaldson[7,8,***], *Corinne* Donzaud[10], *Jean-Eric* Ducret[1], *Clarisse* Hamadache[2], *Fairouz* Hammache[2], *Pete* Jones[6], *Marco* La Cognata[6], *Adrien* Laviron[2], *Marek* Lewitowicz[1], *Kgashane* Malatji[8], *Antonio* Massara[6], *Cyril* Pitrou[11,12], *Rosario Gianluca* Pizzone[6], *Giovanni Luca* Guardo[6], *Marek* Płoszajczak[1], *Giuseppe* Rapisarda[5,6], *Bernadette* Rebeiro[9,14], *Brigitte* Roussière[2], *Domenico* Santonocito[6], *Nicolas* de Séréville[2], *Maria Letizia* Sergi[5,6], *Eva* Simeckova[4], *Olivier* Sorlin[1], *Christelle* Stodel[1], *Vincent* Tatischeff[2], *Jean-Charles* Thomas[1], and *Aurora* Tumino.[6,13]

[1]Grand Accélérateur National d'Ions Lourds (GANIL), CEA/DRF-CNRS/IN2P3, 14076 Caen Cedex 05, France.
[2]Université Paris-Saclay, CNRS/IN2P3, IJCLab, 91405, Orsay, France.
[3]School of Physics and Astronomy, University of Minnesota, Minneapolis, Minnesota 55455, USA.
[4]Nuclear Physics Institute ASCR, 250 68 Řež, Czech Republic.
[5]Dipartimento di Fisica e Astronomia "E. Majorana", Univ. di Catania, Italy.
[6]Laboratori Nazionali del Sud, INFN, Catania, 95123 Italy.
[7]School of Physics, University of the Witwatersrand, Johannesburg 2050, South Africa.
[8]SSC Laboratory, iThemba LABS, Somerset West 7129, South Africa.
[9]Univ. Lyon, Univ. Claude Bernard Lyon 1, CNRS/IN2P3, IP2I Lyon, UMR 5822, F-69622, Villeurbanne, France.
[10]APC, Université Paris Diderot, CNRS/IN2P3, CEA/IRFU, Observatoire de Paris, Sorbonne Paris Cité, 75205 Paris, France.
[11]Institut d'Astrophysique de Paris, CNRS UMR 7095, 98 bis Bd Arago, 75014 Paris, France.
[12]Sorbonne Université, Institut Lagrange de Paris, 98 bis Bd Arago, 75014 Paris, France.
[13]Facoltà di Ingegneria ed Architettura, Kore University, Viale delle Olimpiadi, 1, I-94100 Enna, Italy.
[14]Department of Physics, McGill University, Montréal, Quebec H3A 2T8, Canada.

Abstract. Several experiments around the world are looking for a new particle, named Dark Boson, which may do the link between the Ordinary Matter (which forms basically stars, planets, interstellar gas...) and the Hidden Sectors of the Universe. This particle, if it exists, would act as the messenger of a new fundamental interaction of nature. In this paper, the underlying Dark Sectors theory will be introduced first. A non-exhaustive summary of experimental studies carried out to date and foreseen in the incoming years will be presented after, including the ^{8}Be anomaly. The last section will provide a status of the New

*New JEDI scientific coordinator / e-mail : bastin@ganil.fr
**New JEDI co-spokesperson
***FASERED co-spokesperson

JEDI**** project which aims to investigate the existence or not of a Dark Boson in the MeV range.

1 The Dark Sectors theory as a new alternative scenario to the standard direct Dark Matter detection approach

Understanding the composition and functioning of our Universe are among the most fundamental and challenging questions in Physics. Despite its apparent complexity, ordinary matter (OM) is in reality composed of just a few elementary particles (quarks and leptons). These particles, together with the force carrying and the Higgs Bosons constitute the main components of the Standard Model of Particle Physics [1]. This theory is able to describe 5% of the Universe (made of OM). 26% of the Universe is made of Dark Matter that does not emit light and whose existence has been inferred as necessary to explain, e.g., the star's rotational velocity within galaxies or the gravitational lensing effects observed in space. The remaining 69% of the Universe is made of Dark Energy, distributed uniformly, acting as a repulsive force counterbalancing gravity. The existence of Dark Energy could explain the accelerated expansion of the Universe. The relative composition of the Universe was deduced from the Cosmic Microwave Background anisotropy [2], in accordance with the prediction of Jim Peebles, awarded the 2019 Nobel Prize in Physics [3]. This is our current and conventionally accepted description of the structure of the Universe. To date, the real composition of the classical Dark Sectors remains a mystery. We still do not have a clear answer to the question "of what is Dark Matter composed?". There are some indications, several candidates, but, despite all the efforts over the last three decades all around the world [4], none of the dedicated experiments has proved the existence of dark particles. To date, the intrinsic nature of dark matter remains a mystery.

To address this issue, several models have been developed and in recent years particular attention is paid to studies of so-called Dark / Hidden Sectors, introduced in refs. [5][6][7][8] but poorly tested [9]. It is based on the idea that we may also consider an indirect interaction between Ordinary Matter and Dark Sectors of the Universe. In this framework, Dark Sectors can be defined as hypothetical sets of relatively light particles with interaction orders of magnitude lower than the electromagnetic interactions. These Dark Sectors may or not include Dark Matter particles. We will keep this definition of Dark Sectors in the following text. From a theoretical point of view, Dark Sectors are composed of one or more particles, that couple to the Visible Sector - Ordinary Matter well described by the Standard Model - through portals, as described in Fig. 1. A Dark Boson will act as the messenger of this interaction. It is the Dark Force carrier. This leads to the fundamental question of the existence of a fifth force of nature (in addition to the gravitational, electromagnetic, strong and weak forces). Different portals could be possible, depending on the quantum nature of the Dark Boson (spin and parity). Nuclear Physics experiments allow the study of scalar, pseudo-scalar, vector and pseudo-vector hypothesis.

Non-trivial dynamics in the hidden sectors model can lead to novel phenomenology, whose thorough characterization has only begun (see next section).

2 Dark Sectors experimental studies : state-of-art and ^{8}Be anomaly

Currently, there is much activity in both North America and Europe, evaluating various proposals attempting to see how different experiments can better probe Dark Sectors [9][10]. In

****acronym of New Judicious Experiments for Dark sectors Investigations

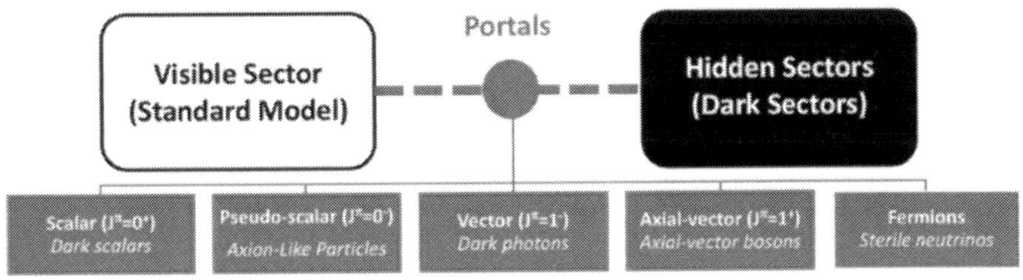

Figure 1. Illustration of the Standard Model coupling to Dark Sectors through given portals. Portals that might be studied though nuclear physics experiments are indicated with a red frame.

the US, experiments at Jefferson lab, SLAC and Fermilab facilities have been discussed. In 2016, CERN launched the "Physics Beyond Colliders" study group to assess the possibility of using the CERN facility for physics not necessarily related to the LHC. One of the physics goals of these exercises is to look at the sensitivity of various proposals for the study of Dark Sectors.

From this prospective works, one may see that, up to now, many experiments using different techniques have just provided exclusion zones, ruling out certain masses and interaction strengths of the possible Dark Boson. An example is given in Fig. 2 of [11] concerning the Dark Photon (vector) portal hypothesis (see Fig. 1). Until now, the scientific community that conducts experimental studies on Dark Sectors is mainly the High Energy Physics one. However, during the last years, there is a Nuclear Physics experiment that has caught the scientific community's attention, performed by a Hungarian research team [12]. Our investigation is further motivated by the recent claim that the anomaly they observed in the electron-positron pair decays of excited states in ^{8}Be can be interpreted as the signature of a hypothetical Dark Boson : a bump appears when looking at the e+ e- angular correlation spectrum. It has been interpreted as the possible signature of a Dark Boson decay, which was named X17. This experiment provides the main constraint on the Dark Boson mass : m_X=16.70 ± 0.35(stat) ± 0.5(syst) MeV/c^2. From this experiment, the following branching ratios were also extracted: proton decay, γ-decay, internal pair creation and ejection of a new (X17) particle. Has a Dark Boson finally been observed? It is not very clear yet. First, the given result is not compatible with the conclusions from the NA48 complementary experiment performed at CERN [13]. Compatibility might be enforced if some extreme hypothesis are considered, such as a protophobic dark photon [11]. But, the ATOMKI group has also later studied the signature of a Dark Boson in ^{4}He and ^{12}C, claiming the existence of a similar anomaly [14][15][16]. However, the synthesis made by J.L. Feng *et al.* [17] underlines (see Tab. III of the article) very well that it is difficult to build up a coherent conclusion concerning the quantum nature of the X17 particle based on all the ATOMKI experimental results. A possible explanation could be a X17 production mechanism rather proceeding through a direct capture process than through any nuclear resonance, as suggested by ref. [18]. Therefore, an independent measurement is needed.

There are lots of experimental programs that started recently to study the X17 anomaly. Even if they look like providing very promising sensitive probes, one may see that experiments conventionally used for high energy physics such as LHCb need major updates and could provide a first experimental run very later on (final detector expected 2025) [19]. NA64 managed to realize an additional experimental run in 2018. The results obtained did not allowed to conclude about the X17 anomaly but allow to improve the sensitivities of their previous measurement [20] and will need an upgrade of the setup to go further on this topic [21]. PADME, as for it, could search for this dark photon with a few proper experimental set-up changes [22]. In parallel, new emerging Nuclear Physics experimental programs, including New JEDI, are under development or reached the commissioning phase, even some have carried out already a first experiment [23][24][25][26][27]. The diversity of the detection technologies and reaction mechanisms envisaged for all these experimental programs should provide, in the near future, an unambiguous clarification on the existence or not of the hypothetical X17 particle. This joined effort of the Nuclear Physics community should allow to investigate, in a general manner, the existence of any Dark Boson in the MeV *terra incognita*. In the next section, we will provide details concerning the New JEDI project, on the way to realize its first long experimental run on ^{8}Be at the ANDROMEDE facility [28].

3 The New JEDI project : presentation and status

The New JEDI project aims to provide, in a general manner, quick and reliable tests of the existence of a light Dark Boson (MeV range) that couples the Standard Model to Dark Sectors. Fig. 2 shows the mechanical scheme (at the center) of the New JEDI setup first version and pictures of its components (chamber, detectors, target holder and units). This version has been used during the commissioning that took place in June 2020 at the tandetron facility located at Řež, in Czech Republic [29]. It was developed based on test experiments conducted at the ARAMIS facility (Orsay, France) [30] and Geant 3 simulations, carried out since 2017.

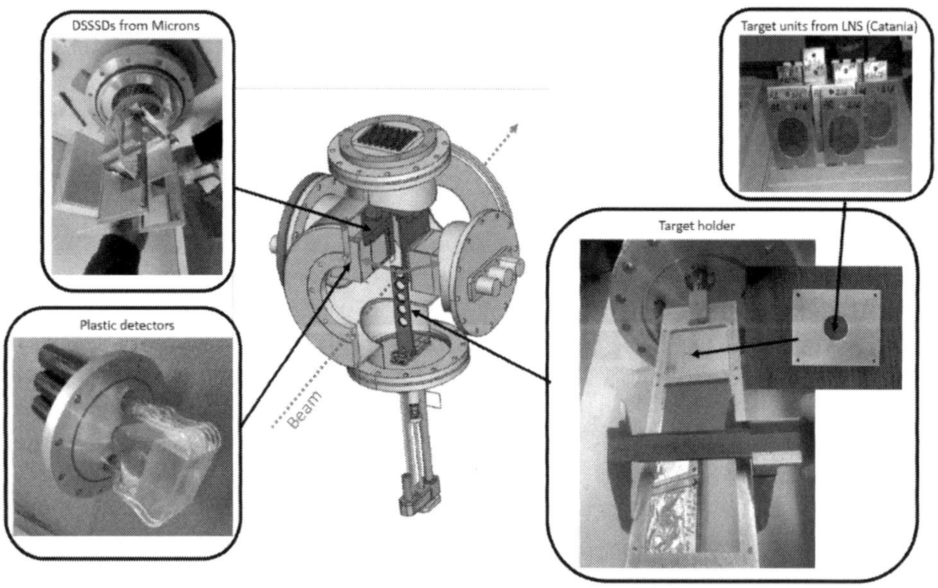

Figure 2. Mechanical scheme (at the center) of the New JEDI setup (first version) and pictures of its components (chamber, detectors, target holder and units).

The set of Double-Sided Silicon Strips Detectors (DSSSDs) of the New JEDI setup provides energy losses and angles of the detected electrons and positrons. In addition, the sets of plastic detectors is used to measure the residual energy of electrons and positrons. They are also

used to veto external background events. The detection system is coupled to the new generation NUMEXO2 digitizers [31]. A specific charge integration firmware and a coincidence cross-check mode have been developed for the project. It is worth noting that the New JEDI detection system appears to be close to the new version developed in parallel by the ATOMKI group. The geometry chosen here is more focused on the detection of X17-like events.
For the commissioning, we used mainly a proton beam of 470 keV (also 1070 keV) impinging onto a LiF target of typically 150 μgr/cm^2 thickness, having a carbon backing with a thickness of 30 μgr/cm^2. The conclusions from the commissioning are the following: there is a major contribution of the ^{19}F(p,α)^{15}O reaction in the DSSSDs that needs to be stopped using a thick (aluminized mylar) foil in front of the detector. The Lithium target emits also a lot of electrons that can be stopped in this front foil providing it is a conductive one, to avoid charge and discharge effects. Surprisingly, there is also a need to protect the back of the DSSSDs as well which see also electrons (from back-scatterings with plastic detectors...). We observed multi-scattering processes that need to be investigated more in details in the next experiment. From simulations, we identified that the electron-electron and electron-nucleon interactions play an important role (same behavior for positrons). The methods developed to measure and to check the proper coincidences between the different detectors seem to work properly. Moreover, we studied the behaviour of CaF and PTFE targets developed by the INFN-LNS laboratory, part of the New JEDI collaboration. They can be useful to disentangle the contribution of reaction from F in the spectra. They showed very promising characteristics online. We realized that we need to refine the optimal target thicknesses to get benefit of the full beam intensity expected from the simulations. In this configuration, a reasonable efficiency of 10% is estimated for the X boson electron-positron pair decay, if it exists. We also observed interesting geometrical effects (e.g. shadow effects in the DSSSDs and from the target ladder). All this will be investigated and optimized for the first experimental run foreseen at the ANDROMEDE facility.
We plan to develop a long-term research program in the MeV *terra incognita* energy range at the new SPIRAL2 facility (Caen, France) [32], that will deliver unique high-intensity beams of light, heavy-ions and neutrons in Europe. In practice, three experiments using the New JEDI setup concerning the existence or not of the X17 Dark Boson are envisaged between 2022 and 2025: two at the ANDROMEDE facility and one at SPIRAL2 looking for the signature in ^{8}Be and ^{3}He quantum systems, respectively. To start off, we plan for example to populate "excited" non-resonant state in ^{3}He around 18 MeV using the high-power pulsed proton beam of the LINAC impinging onto a thin CD2 target. In addition, one crucial experiment (not using the New JEDI detector) focused on ^{8}Be spectroscopic decay study is foreseen at iThemba LABS using the unique coupling of low-energy beam from the tandetron accelerator and the half-AFRODITE multi-detectors array (FASERED project). The aim is to investigate deeply if the X17 signal can be an artefact linked to an experimental error or a subtle nuclear physics effect in ^{8}Be, which has a complex nuclear structure.

Acknowledgments

The authors would like to address a special thank you to the GANIL DELTA group for their precious contribution on the electronics and data acquisition developments, as well as the GANIL and IJCLab mechanical design groups and the NPI and tandetron technical staffs for their outstanding work and support to the New JEDI project. Authors also acknowledge prof. S. Mirabella and collaborators for the CaF targets characterization at the 3MV Singletron of DFA-UniCT. This work has been supported by the French-Czech LEA NuAG collaboration. We acknowledge the support of the MEYS project EF16-013/0001679. For their support via the projet PHC PROTEA - FASERED - 47796WF, the authors would like also to acknowledge the French Ministry of Foreign Affairs and the Ministry of Higher Education and Research from the French side, and the Department of Science and Innovation (DSI) and the National Research Foundation (NRF) in South Africa.

References

[1] R.L. Workman *et al.* (Particle Data Group), Prog. Theor. Exp. Phys. 2022, 083C01 (2022).
[2] W. Hu and S. Dodelson, Annu. Rev. Astron. Astrophys. **40**, 171 (2002).
[3] P.J.E. Peebles, Rev. Mod. Phys. **92**, 030501 (2020).
[4] https://en.wikipedia.org/wiki/Category:Experiments_for_dark_matter_search
[5] C. Boehm *et al.*, NPB **683**, 219 (2004).
[6] M. Pospelov *et al.*, Phys. Lett. B **662**, 53 (2008).
[7] N. Arkani-Hame *et al.*, Phys. Rev. D **79**, 015014 (2009).
[8] G. Lanfranchi *et al.*, Annu. Rev. Nucl. Part. Sci. **71**, 279 (2021).
[9] J. Alexander *et al.*, arXiv:1608.08632.
[10] M. Battaglieri *et al.*, arXiv:1707.04591.
[11] J. L. Feng *et al.*, Phys. Rev. Lett. **117**, 071803 (2016).
[12] A. J. Krasznahorkay *et al.*, Phys. Rev. Lett. **116**, 042501 (2016).
[13] J. R. Batley *et al.* (NA48/2 Collaboration), Phys. Lett. B **746**, 178 (2015).
[14] A.J. Krasznahorkay *et al.*, Acta Phys. Pol,. B **50**, 675 (2019).
[15] A.J. Krasznahorkay *et al.*, Phys. Rev. C **104**, 044003 (2021).
[16] A.J. Krasznahorkay *et al.*, arXiv:2209.10795.
[17] J.L. Feng *et al.*, Phys. Rev. D **102**, 036016 (2021).
[18] X. Zhan and G. A. Miller, Phys. Lett. B **813**, 136061 (2021).
[19] Cesar L. da Silva *et al.*, XXVI Cracow EPIPHANY Conference, LHC Physics: Standard Model and Beyond, January 2020 : https://cds.cern.ch/record/2706116/files/Epiphany-Cesar%2008.01.pdf
[20] D. Banerjee *et al.* (NA64 Collaboration), Phys. Rev. D **101**, 071101 (2020).
[21] D. Banerjee *et al.* (NA64 Collaboration), Eur. Phys. J. C **80**, 1159 (2020).
[22] E. Nardi *et al.*, Phys. Rev. D **97**, 095004 (2018).
[23] M. Brodeur and K.G. Leach, contribution to "U.S. Cosmic Visions: New Ideas in Dark Matter 2017".
[24] R. F. Lang *et al.*, "8BeP: A 8Be Experiment at Purdue", U.S. Cosmic Visions: New Ideas in Dark Matter 2017.
[25] H. Natal da Luz *et al.*, RD51 Collaboration Meeting and topical workshop on "New Horizons in TPCs", October 2020.
[26] W. Mittig *et al.*, NSCL, The green sheet Vol. **41** , Num. 30 (2020).
[27] C. Gustavino *et al.*, EURO-LABS contribution 2021.
[28] https://www.ijclab.in2p3.fr/en/platforms/andromede/
[29] http://www.ujf.cas.cz/en/departments/department-of-neutron-physics/lnam/
[30] https://jannus-scalp.ijclab.in2p3.fr/en/irradiations/aramis/
[31] C. Houarner *et al.*, in preparation.
[32] https://www.ganil-spiral2.eu/
[33] https://tlabs.ac.za/

Trojan Horse Method: a general introduction

Stefano Romano[1,2,3,*] for the ASFIN collaboration

[1]INFN-LNS, Catania, Italy,
[2]Dipartimento di Fisica e Astronomia, Universitá di Catania, Catania, Italy
[3]Centro Siciliano Fisica Nucleare, Catania, Italy

Abstract. Owing the presence of the Coulomb barrier at astrophysically relevant kinetic energies, it is very difficult, or sometimes impossible to measure astrophysical reaction cross sections in laboratories, especially for the presence of the electron screening effect. This is why different indirect techniques are being used along with direct measurements. The Trojan Horse Method (THM) is a unique indirect technique allowing one to measure astrophysical rearrangement reactions down to astrophysical relevant energies. The basic principle and a review of the main applications of the THM are presented.

1 Introduction

Nuclear fusion reactions, that take place in the hot interiors of remote and long-vanished stars over billions of years, are the origin of nearly all the chemical elements in the universe [1]. The detailed understanding of the origin of the chemical elements and their isotopes has combined astrophysics and nuclear physics, and forms what is called nuclear astrophysics. In turn, nuclear reactions are the heart of nuclear astrophysics: they strongly influence the nucleosynthesis in the earliest stages of the universe and in all the objects formed thereafter, and control the associated energy generation (by processes called nuclear fusion or nuclear burning), neutrino luminosity, and evolution of stars. A good knowledge of the rates of these fusion reactions is essential for understanding this broad picture

In a stellar plasma the constituent nuclei are usually in thermal equilibrium at some local temperature T. Occasionally they collide with other nuclei, whereby two different nuclei can emerge from a collision A+x→ c+C. The cross section σ(E) of nuclear rearrangement reaction A(x,c)C is of course governed by the laws of quantum mechanics where, in most cases, the Coulomb and centrifugal barriers arising from nuclear charges and angular momenta in the entrance channel of the reaction strongly inhibit the penetration of one nucleus into another. This barrier penetration leads to a steep energy dependence of the cross section. It is the challenge to the experimentalist to make precise σ(E) measurements at the Gamow energy (E_G), i.e. the energy range where the reaction rate in a given temperature range is maximum. Owing to the strong Coulomb suppression, the behavior of the cross section at E_G is usually extrapolated from the higher energies by using the definition of the smoother bare nucleus astrophysical factor S_b(E).

Although the S_b(E)-factor allows for an easier extrapolation, large uncertainties to $\sigma(E_G)$ may be introduced due to for instance the presence of unexpected resonances, or high energy

*e-mail: romano@lns.infn.it

tails of sub-threshold resonances. In order to avoid the extrapolation procedure, a number of experimental solutions were proposed in direct measurements for enhancing the signal-to-noise ratio at E_G.
Moreover, the measurements in the laboratory at ultra-low energies suffer from the complication due to the effects of electron screening [2]. This leads to an exponential increase of the laboratory measured cross section $\sigma_s(E)$ [or equivalently of the astrophysical factor $S_s(E)$] with decreasing energy relative to the case of bare nuclei. Then, although it is possible to measure cross sections in the Gamow energy range, the bare nucleus cross section σ_b is extracted by extrapolating the direct data behavior at higher energies where negligible electron screening contribution is expected. In order to decrease uncertainties in the case of charged particle induced reactions, extrapolations should be avoided and therefore experimental techniques were improved (e.g. by means of underground laboratories). After improving measurements (at very low energies), electron screening effects were discovered. Finally to extract from direct (shielded) measurements the bare astrophysical S_b(E)-factor, extrapolations were performed at higher energy. In any case the extrapolation procedure is once more necessary.

2 The Trojan Horse Method

Alternative methods for determining bare nucleus cross sections of astrophysical interest are needed. In this context a number of indirect methods, e.g. the Coulomb dissociation (CD) [3], the Asymptotic Normalization Coefficient method (ANC)[4, 5] and references therein and the Trojan-horse method (THM) were developed [6, 7]. For further information on the development and first principles of the method please refer to [8, 9]. The latter has already been applied several times to reactions connected with fundamental astrophysical problems such as primordial nucleosynthesis [10–13], lithium problem [14–16, 28], light elements depletion [20, 24], AGB [39] and Novae nucleosynthesis [33]. It was also applied to reactions induced by radioactive ion beams [34, 35] and neutrons [37, 47]. THM selects the quasi-free (QF) contribution of an appropriate three-body reaction performed at energies well above the Coulomb barrier to extract a charged particle two-body cross section at energies of astrophysical interest. The idea of the THM [7] is to extract the cross section of an astrophysically relevant two-body reaction

$$A + x \rightarrow c + C \tag{1}$$

at low energies from a suitable chosen three-body quasi-free reaction

$$A + a \rightarrow c + C + S \tag{2}$$

In this approach S acts as a spectator to the $A + x \rightarrow c + C$ binary interaction.This is done with the help of direct processes theory assuming that the Trojan Horse nucleus a has a strong $x \oplus S$ cluster structure [19, 45]. In many applications, this assumption is trivially fulfilled e.g. a = deuteron, x = proton, S = neutron. If the bombarding energy E_A is chosen high enough to overcome the Coulomb barrier in the entrance channel of the three-body reaction, both Coulomb barrier and electron screening effects are negligible. The polar approximation, used in the standard THM prescription has been extensively verified [40, 41] and constitutes a powerful validity test for the method which strenghtens the theoretical approach. We refer to [17, 18] for further and advanced theoretical approach to the method.

We just underline that THM allows to link the three-body cross section which is measured in the laboratory with the half-off-energy shell cross section of the binary process of astrophysical interest. Then after inclusion of the Coulombian effects data are then compared and normalized to direct data, at the higher energies available. After that phase, the reaction rate is calculated according to the standard prescriptions.

2.1 Identification and selection of the three-body reaction of interest

The first step in data-analysis is to identify the events related to the three-body reaction of interest for THM, $A + a \to c + C + S$ from the other ones occurring in the target. This is accomplished by studying the kinematic locus related to the above reaction and the Q-value spectrum. Coincidences between detectors aiming at c and C are examined and a typical plot of the particle energy detected in each detector is performed (see for instance fig. 4 of [40]). A narrow angular range ($\approx \pm 2^o$) is selected on both detectors and events coming from an appropriate Monte Carlo simulation, taking into account the geometrical properties of the experimental set-up as well as the features of the detectors. If a good agreement shows up, this allows further studies. Using a graphical cut which selects only the events overlapping with the Monte Carlo simulation, the Q-value spectrum is plotted (see figure 6 in [42]). A peak compatible, within the experimental errors, with the theoretical Q-value of the three-body reaction is expected. From now on only the events below the Q-value peak and inside the kinematical graphical cut in the kinematic locus will be used for further data-analysis, in order to be relatively confident that the reaction channel $A + a \to c + C + S$ is selected.

2.2 Selection of the quasi-free mechanism

The next step after identifying the 3-body process is to investigate the reaction mechanisms involved and to separate the quasi-free (QF) contribution from any other kind of reaction mechanisms as required by the THM prescriptions. This can be done by studying, among all the available observables, the most sensitive to the reaction mechanisms which is, by far, the shape of the momentum distribution, $|\varphi(\mathbf{p_s})|^2$. According to the prescriptions in [43, 44], the momentum distribution of the third and un-detected particle is examined. This gives a major constraint for the presence of the QF mechanism and the possible application of the THM. In order to extract the experimental momentum distribution of the the undetected particle, (the spectator after the QF process is identified and selected) , $|\varphi(\mathbf{p_s})|^2_{exp}$, the energy sharing method can be applied to each pair of coincidence detectors, selecting energy intervals, ΔE_{cm}. Keeping in mind the factorization of Eq. (3), see below, since $[(d\sigma/d\Omega)_{cm}]^{HOES}$ is nearly constant in an adequate energy interval, one can get the shape of the momentum distribution of the undetected neutron directly from the coincidence yield divided by the kinematical factor, as calculated from a suitable Monte Carlo simulation. The obtained typical momentum distribution is reported for instance in [44]. It is also compared with the theoretical distribution calculated from the Hulthen function (dashed line) with parameters taken from [45]. We can see how within the experimental errors the theoretical curve reproduces the experimental data, thus confirming the hypothesis that the neutron is acting as a spectator and that the process under investigation is a quasi-free mechanism. We only considered the s-wave since other contribution, i.e. the d-wave, were shown to be negligible [45]. According to the prescription adopted in [9] and in the standard THM approaches, only data in a limited $|p_s|$ range were chosen according to the chosen Trojan horse nucleus and used in the further analysis.

Once the experimental full width at half maximum (FWHM) Γ is obtained it has to be compared with the asymptotic theoretical value (e.g. for deuteron about 58 MeV/c) in order to highlight the presence of possible distortions. If those are present they should be taken into account as reported in [43, 44]. After this test we can stress the role of the spectator to the QF process, which constitutes a solid base for the further THM application to the studied three-body reaction for retrieving information on the binary reaction bare nucleus cross section of interest at astrophysical energies. For the following analysis only data with p_s <30 MeV/c are considered as arising from quasi-free mechanisms. This, of course, depends on the chosen

Trojan Horse nucleus; the above value of p_s <30 MeV/c is typical in most cases for a l =0 relative inter cluster motion as for ^{2}H, ^{6}Li, ^{3}He.

Reaction mechanisms other than the quasi-free one should be considered as a "noise" and removed from the dataset eligible for THM application. This is done for example for the sequential mechanisms, usually occurring in the target, where an intermediate compound nucleus is formed. Examples are shown in [28]. Once the quasi-free mechanism is selected the bulk of data available for THM analysis is consistently reduced. For typical cases around 90% of data are usually rejected because they are not arising from a quasi-free process.

2.3 Extraction of the binary reaction cross section

In the standard THM analysis, the two body cross section is derived by dividing the experimental three-body one by the product of the kinematic factor modulated by the momentum distribution of the spectator inside the Trojan Horse nucleus [8], i.e.

$$\left(\frac{\mathrm{d}\sigma}{\mathrm{d}\Omega}\right)^{\mathrm{HOES}} \propto \frac{\mathrm{d}^3\sigma}{\mathrm{dE}_{\alpha_1}\mathrm{d}\Omega_{\alpha_1}\mathrm{d}\Omega_{\alpha_2}} / \left(\mathrm{KF}\cdot|\varphi_{\mathrm{exp}}(\mathbf{p_s})|^2\right) \tag{3}$$

Usually the factors $KF \cdot |\varphi_{\mathrm{exp}}(\mathbf{p_s})|^2$ are calculated by means of a Monte Carlo simulation, taking into account the geometrical position of the detectors. The width of the momentum distribution is set to the experimentally measured value in order to account for the distortion effects arising at low transferred momenta as discussed in [43].

The extracted $[d\sigma/d\Omega]^{\mathrm{HOES}}$ as a function of E_{cm}, corrected for the penetration factor (usually described in terms of the Regular and Irregular Coulomb functions) has to be compared to direct data in order to perform normalization at higher available energy. This allows to calculate and compare also the astrophysical S(E)-factor, which is extracted from the cross section according to the usual definition. This has been done in a big number of cases, many times both above and below the Coulomb barrier thus showing the strength of the method. After this necessary step the reaction-rate can be calculated according to the standard expression available in literature. Multiple examples of this procedure and some results are given here for reference [22].

Despite several outstanding achievements like the application to radioactive ion beams or to neutron emitting or neutron induced reactions ([36, 50]), the method has still some points that should be improved, like for the necessity of normalization and the possibility of absolute measurements, the possible application to (p,γ) and (α, γ) reactions and so on. The possibility of measuring absolute cross sections is, for example, under consideration for the next applications in order to avoid the normalization procedure to direct data.

A summary of the recent results obtained by means of the THM method is summarized in table 1 together with references. Table 1 shows success for (p, α), (α, p), (α, n), (p, n), (n, α) and (n, p) reactions of astrophysical interest studied with the THM method, both with stable as well unstable beams. Moreover, our study of ^{12}C(^{12}C,α)^{20}Ne and ^{12}C(^{12}C,p)^{23}Na [27] has demonstrated that we can extend the method to heavier ions [32].

The THM is complementary to direct measurements which are needed in the higher energy range where indirect data must be normalized to available data in literature. A numbers of astrophysical scenarios were thus explored helping to shed light in many cosmic problems. We underline that in many cases the THM will be a powerful aid to nuclear astrophysics knowledge especially in the investigation of the interaction of neutron-induced reactions with unstable particles.

References

[1] E.M.Burbidge, G.R.Burbidge, W.A.Fowler and F.Hoyle, Rev. Mod. Phys. **29**, (1957) 547
[2] H.J.Assenbaum, K.Langanke and C.Rolfs, Z. Phys.A **327**, (1987) 461
[3] G.Baur, C.A. Bertulani & H Rebel , Nucl. Phys. A **458**, (1986) 188
[4] A.M.Mukhamedzhanov and R.E.Tribble, Phys.Rev. C **59**, (1999) 3418
[5] G.G. Kiss, M. La Cognata, C. Spitaleri et al., (2020) Phys. Lett. B, 135606
[6] G.Baur, Phys. Lett. B **178**, (1986) 135
[7] C.Spitaleri 1991, in *Problems of Fundamental Modern Physics, II* : Proceedings, Ed. by R. Cherubini,P. Dalpiaz, and B. Minetti (World Sci., 1991), p. 21
[8] Spitaleri, C., La Cognata, M., Lamia, L., Mukhamedzhanov, A. M., & Pizzone, R. G. 2016, European Physical Journal A, **52**, 77
[9] Spitaleri, C.; La Cognata, M.; Lamia, L.; Pizzone, R. G.; Tumino, A.; *Eur. Phys. Journ. A* **2019**, *55*, 161.
[10] A. Rinollo et al., Nucl. Phys. A **758**, (2005)146c
[11] A. Tumino et al., Phys. Lett. B **705**, (2011) 545
[12] R.G. Pizzone et al., Astrophys. J. **786**, (2014) 112
[13] C. Li, et al., Phys. Rev. C **92** (2015) 025805
[14] Aliotta, M., Spitaleri, C., Lattuada, M., et al. 2000, European Physical Journal A, **9**, 435
[15] R.G. Pizzone et al., A. & A. **438**, (2005) 779
[16] L. Lamia et al., Astrophys. J. **768**, (2013) 65
[17] A. Tumino, et al., Ann. Rev. Nucl. Part. Sc. **71**, 345 (2021)
[18] A. Tumino et al., in: Handbook of Nuclear Physics (Springer, Singapore, 2022)
[19] R.G. Pizzone, C. Spampinato et al., Eur. Phys. J. , **A56**, 199 (2020)
[20] C. Spitaleri et al., Phys. Rev. C **95**, (2017) 035801
[21] A. Caciolli et al., Eur. Phys. Jour. A **52**, 136, 2016
[22] S. Romano, et al., Eur. Phys. Jour. A **27**, (2006) 221
[23] Wen, Q.-G., Li, C.-B., Zhou, S.-H., et al. 2008, Phys. Rev. C **78**, 035805
[24] L. Lamia et al., Astrophys. J. **811**, (2015) 99
[25] M. La Cognata et al., Eur. Phys. J. A**27**, (2006) 249
[26] La Cognata, M., Spitaleri, C., Mukhamedzhanov, A., et al. 2010, The Astrophysical Journal, **708**, 796
[27] Indelicato, I., La Cognata, M., Spitaleri, C., et al. 2017, The Astrophysical Journal, **845**, 19
[28] L.Lamia, et al., J. Phys. G 39, (2012) 015106
[29] Pizzone, R. G., D'Agata, G., La Cognata, M., et al. 2017, The Astrophysical Journal, **836**, 57
[30] M.Lattuada et al., Ap. J. **562**, (2001) 1076
[31] D'Agata, G.; Pizzone, R. G. et al. 2018, The Astrophysical Journal, **860**, 61
[32] A. Tumino, C. Spitaleri, M. La Cognata, S. Cherubini, G. L. Guardo, M. Gulino, S. Hayakawa, I. Indelicato, L. Lamia, H. Petrascu, R. G. Pizzone et al., 2018, Nature, **557**, 687-690
[33] Sergi, M. L., Spitaleri, C., La Cognata, M., et al. 2015, Phys. Rev. C, **91**, 065803
[34] S. Cherubini et al., Phys. Rev. **C 92**, (2015) 015805
[35] Pizzone, R. G., Roeder, B. T., McCleskey, M., et al. 2016, European Physical Journal A, **52**, 24

[36] La Cognata, M., Pizzone, R. G., José, J., et al. 2017, The Astrophysical Journal, **846**, 65
[37] Guardo, G. L., Spitaleri, C., Lamia, L., et al. 2017, Phys. Rev. C, **95**, 025807
[38] La Cognata, M.; Spitaleri, C.; Trippella, O.; et al., 2012, Phys. Rev. Lett.,**109**, 232701
[39] S. Palmerini et al., Ap. J., **764**, (2013) 128
[40] R.G. Pizzone et al. 2011, Phys. Rev. C, **83**, 045801
[41] R.G. Pizzone et al. 2013, Phys. Rev. C **87**, 025805
[42] A.Tumino et al. 2008,Phys. Rev. C **78**, (2008) 064001
[43] R.G. Pizzone et al., Phys. Rev. C **71**, (2005) 058801
[44] R.G. Pizzone et al., Phys. Rev. C **80**, (2009) 025807
[45] L.Lamia,M. La Cognata, C. Spitaleri, B. Irgaziev, R.G. Pizzone, Phys. Rev. C **85**, (2012) 025805
[46] Pizzone R.G. et al., Eur. Phys. J. A, **56** 283 (2020)
[47] M. Gulino et al., Phys. Rev. C **87**, (2013) 012801
[48] La Cognata, et al., 2005, Phys. Rev. C, **72**, 065802
[49] L. Lamia et al., Ap. J. **850**, 175, (2017)
[50] L. Lamia, M. Mazzocco, R.G. Pizzone, et al. , Astrophys. J., **879**, 23 (2019)
[51] S.Hayakawa, M. La Cognata, L. Lamia et al., Astroph. J,**915**, L13, (2021)
[52] S. Palmerini, M. La Cognata, F. Hammache et al., 2021, European Physical Journal A, **136**, 898
[53] M. La Cognata, S. Palmerini, P. Adsley et al., 2022, Astrophys. J., **941**, 96

Table 1. Two-body reactions recently studied via Trojan Horse Method at the astrophysical energies.

Binary reaction	Indirect reaction	TH nucleus	Reference
^{7}Li(p,α)^{4}He	^{7}Li(d,α α)n	d = (p⊕n)	[16, 30]
^{6}Li(d,α)^{4}He	^{6}Li(^{6}Li,α α)^{4}H	^{6}Li = (α⊕d)	[14]
^{6}Li(p,α)^{3}He	^{6}Li(d,α ^{3}He)n	d = (p⊕n)	[28]
^{11}B(p ,α)^{8}Be	^{11}B(d ,^{8}Be α)n	d = (p⊕n)	[28]
^{10}B(p,α)^{7}Be	^{10}B(d,^{7}Be α)n	d = (p⊕n)	[20, 21]
^{9}Be(p,α)^{6}Li	^{9}Be(d,^{6}Li α)n	d = (p⊕n)	[22, 23]
^{2}H(^{3}He,p)^{4}He	^{6}Li(^{3}He,p α)^{4}He	^{3}He = (p⊕n)	[25]
^{2}H(d,p)^{3}H	^{2}H(^{6}Li,t p)^{4}He	^{6}Li = (α⊕d)	[10]
^{18}F(p,α)^{15}O	^{18}F(d,α^{15}O)n	d = (p⊕n)	[34–36]
^{15}N(p,α)^{12}C	^{15}N(d,α^{12}C)n	d = (p⊕n)	[25]
^{18}O(p,α)^{15}N	^{18}O(d,α^{15}N)n	d = (p⊕n)	[26]
^{19}F(p,α)^{16}O	^{19}F(d,α^{16}O)n	d = (p⊕n)	[27]
^{19}F(α,p)^{22}Ne	^{19}F(^{6}Li,p^{22}Ne)α	^{6}Li = (α⊕d)	[29, 31]
^{12}C(^{12}C,α)^{20}Ne	^{12}C(^{14}N,α d)^{20}Ne	^{14}N=^{12}(C ⊕d)	[32]
^{17}O(p,α)^{14}N	^{17}O(d,α^{14}N)n	d = (p⊕n)	[33]
^{17}O(n,α)^{14}C	^{17}O(d,α^{14}C)H	d = (p⊕n)	[37, 47]
^{13}C(α,n)^{16}O	^{13}C(^{6}Li,n d)^{16}O	^{6}Li = (α⊕d)	[38]
^{12}C(^{12}C,p)^{23}Na	^{12}C(^{14}N,p d)^{23}Na	^{14}N=^{12}(C ⊕d)	[32]
^{3}He(d,p)^{4}He	^{3}He(^{6}Li,p α)^{4}He	^{6}Li = (α⊕d)	[48]
^{3}He(n,p)^{3}H	^{3}He(^{2}H,pp)^{3}H	^{2}H = (p⊕n)	[19]
^{7}Be(n,α)^{4}He	^{7}Be(^{2}H,$\alpha\alpha$)n	^{2}H = (p⊕n)	[49–51]
^{27}Al(p,α)^{24}Mg	^{27}Al(^{2}H,α^{24}Mg)n	^{2}H = (p⊕n)	[52, 53]

The PANDORA project: a setup for in-plasma β-decay studies in nuclei of astrophysical interest

David Mascali[1*] and *Domenico* Santonocito[1]

[1]INFN – Laboratori Nazionali del Sud, via S. Sofia 62, 95123 Catania, Italy

Abstract. Theoretical predictions as well as experiments performed at Storage Rings have shown that a high degree of ionization can affect the half-life of β-radionuclides. The PANDORA project aims at investigating, for the first time, beta decay rates in a plasma, simulating specific stellar-like conditions. A description of the physics motivations and the experimental setup developed to accomplish this task is given. The physics cases selected for the first experimental campaign are also presented.

1 Introduction

Beta decay rates in stars represent a fundamental ingredient in the study of heavy element production since the attempt to reproduce the elemental abundances in the Universe calls for an accurate knowledge of β-decay half-life in stellar environment and the comprehension of the possible mechanisms affecting its variation. Even though beta decay constant λ has been measured for a wide range of nuclei, a major difference exists between terrestrial and stellar conditions. In fact, stellar nucleosynthesis proceeds in a hot and dense environment which affects the degree of ionization of the atoms involved in the process and could play an important role in altering the beta decay process. Early studies of a possible effect of the environment on the nuclear β decay constant didn't show any evidence of a marked variation, indicating, in the best cases, a change smaller than 0.05% [1]. However, in 1947 two independent papers from E. Segrè [2] and R. Daudel et al. [3] suggested the possibility that beta decay rates could be affected by a change in electron density, at least for low Z nuclei. Such an effect, which could play a significant role in altering beta decay rates in stars, was observed for the first time in the electron capture process of ^{7}Be leading to ^{7}Li. A lifetime variation of 3.5% was measured [1].

An important breakthrough in the field was achieved with the advent of Storage Rings due to the possibility of preserving, for extended periods of time (up to several hours), highly ionized ions and mapping their decay. A new decay mechanism, called "bound state β-decay" in which the electron emitted in the decay process is captured in one of the inner shells of the atom was observed for the first time in highly ionized atoms [4]. In neutral atoms or ions with only a few holes in the atomic levels such decay mechanism is hindered or even blocked because the emitted electron is unable to access to a bound final state but in highly ionized ions the decay becomes possible or, somehow, energetically favored. Experimental results

* Corresponding author : davidmascali@lns.infn.it

showed sizeable changes in beta decay half-life. For example, fully stripped $^{187}Re^{75+}$ ions decayed by 9 orders of magnitude faster than neutral ^{187}Re atoms, which have a half-life of 42 Gyr [5]. This phenomenon can be understood in the framework of Fermi beta decay theory which is fundamental to calculate the effect of stellar conditions on β-decay half-lives [6].

On a terrestrial level three classes of beta decays are studied, β^+, β^- and electron capture, typically from K-shell, while in stellar environment two further decay modes should be taken into account, the bound-state beta decay and the free electron capture. Besides, in star environment, due to the high temperature, atoms can be ionized and ions can be in excited states making the evaluation of the reaction Q-value more complex than the one calculated for neutral atomic species. The additional terms in the Q-value calculation needed to take into account the previously described effects can modify its sign, especially for small Q-values, making the decay process possible, differently from the one evaluated in neutral atoms. The total decay rate, in Fermi's theory, is conventionally written in terms of half-life ($t_{1/2}$) and can be evaluated from the expression for the *ft*-value of the decay given by:

$$f_L(Z',Q)t_{\frac{1}{2}} = \frac{(ln2)2\pi^3\hbar^7}{g^2 m_e^5 c^4 \left|M_{if}^L\right|^2}$$

where $f_L(Z',Q)$ is a dimensionless theoretical factor that depends on the phase space of the lepton waves, the type of transition and the type of decay, Z' is the charge of the daughter nucleus, Q is the reaction Q-value, g is the weak interaction strength, $|M_{if}^L|^2$ is the nuclear matrix element and m_e, c, h are respectively, the mass of electron, the speed of light and the Planck constant. Since *ft*-values are dependent only on the nuclear matrix element, the stronger is the coupling between the states of parent and daughter nuclei the larger is the matrix element. This implies a larger chance of a nuclear transition and consequently a smaller *ft*-value. Fixing the value of the matrix elements, the two quantities $f_L(Z',Q)$ and $t_{1/2}$ are correlated and therefore a variation in $f_L(Z',Q)$, which depends on lepton phase and Q-value, affects the $t_{1/2}$. Since plasma environment modifies the reaction Q-value due to the high degree of ionization of the atoms and to the clouds of charges, both positive and negative, which create perturbations to the atomic levels and impact level populations, the stellar and terrestrial decays can have different $f_L(Z',Q)$ and therefore different $t_{1/2}$ for a fixed type of transition. The seminal theoretical work of Takahashi and Yokoi [7] investigated the in-plasma effects due to the temperature and the electron density on beta decay rates. In particular, it was clearly shown the role played by the temperature (and electron density) in reducing the beta decay half-life, the predicted variation attaining even many orders of magnitude, depending on the physics cases investigated.

Takahashi-Yokoi predictions, together with the first experimental results achieved using Storage Rings, have fundamental implications in the *s*-process nucleosynthesis based on the competition between neutron capture and beta-decay rates. This is particularly true in the so-called branching points as a variation in the decay rate directly affects the population of a specific nucleus and the subsequent reaction process. Therefore, to get a deeper knowledge of the in-plasma beta decay rates, it seems mandatory to face up a new experimental challenge, reproduce in laboratory some stellar-like conditions and measure the predicted variations in β-decay half-life.

2 The PANDORA project

In order to elucidate the possible effects connected to the dependence of the beta decay constant on the charge state distribution (CSD) of the decaying ions which is related to the thermodynamical condition of the environment, a totally new approach was conceived by the PANDORA (Plasmas for Astrophysics Nuclear Decays Observation and Radiation for Archaeometry) project [8]. It is based on the realization of an innovative plasma trap [9] to be built at the INFN - Laboratori Nazionali del Sud, and devoted to measure, for the first time, nuclear β-decay rates in a plasma emulating some stellar-like conditions in terms of CSD. Such an approach is complementary to the one adopted in Storage Ring experiments where a single charge state at a time can be investigated and has the advantage that the results can be directly compared to theoretical models once the plasma conditions (density, temperature, CSD) are measured and kept under control during the experiments.

In stars a plasma, defined as a quasi-neutral gas made of charged particles exhibiting a collective behavior, is gravitationally confined while, to run experiments in laboratory, a magnetic confinement based on Lorentz force is used. A proper shaping of the magnetic field is fundamental to achieve the plasma confinement and at the same time to limit the plasma deconfinement losses. Such an approach is typically used in ion sources where charged particles spiralyze around the field lines and can be trapped for several millisecond before being extracted and injected in the accelerators. In the PANDORA project, the plasma will be confined in the trap using a magnetic field configuration called *minimum-B* where the magnetic field strength has a minimum at the center of the trap and increases both axially and radially [9]. It will be achieved using three superconducting coils for axial confinement generating a tunable magnetic field up to 3 Tesla and a superconducting hexapole for radial confinement as shown in fig.1a. The ideal confinement also requires some stringent conditions on plasma equilibrium and stability. These aspects can be investigated looking into the equilibrium between the plasma kinetic pressure and the magnetic confining field pressure: a magnetically stable condition is reached when the plasma kinetic pressure is much smaller than the magnetic pressure.

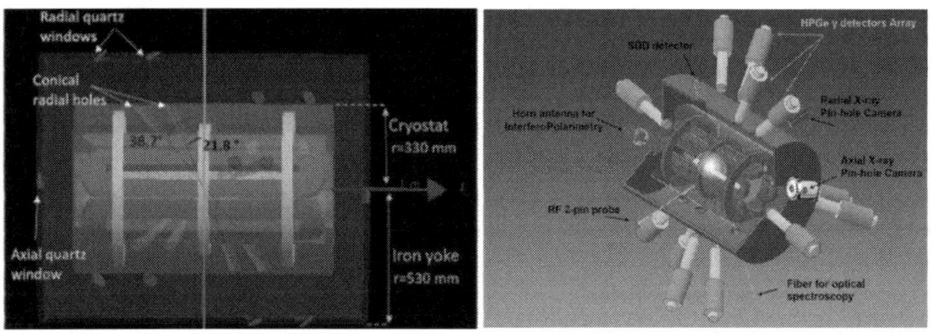

Fig.1 a) Drawing of the PANDORA magnetic system made of 3 superconducting coils (in yellow) for axial confinement and a superconducting hexapole (in red) for radial confinement. The magnetic system is contained in a single cryostat shown in light grey. An iron yoke is used to confine the magnetic field (dark blue area). Conical line of sight are created in the yoke and the cryostat to have direct access to the plasma chamber and used for HPGE detection system and diagnostics. b) Drawing of the PANDORA setup with HPGe around the trap and some diagnostic tools.

Plasma heating is achieved through the interaction of the plasma electrons and electromagnetic waves in the range of few GigaHertz and few kW of power injected in the plasma. The optimum condition for the energy transfer is achieved in the so-called electron-cyclotron resonance region for electromagnetic frequencies equal to the cyclotron frequency given by the relation $\omega_e = eB/m$ where e is the particle charge, B is the magnetic field and m is the particle mass. Ions with high charge states are primarily produced by sequential ionization. In PANDORA they will remain cold, the average temperature (expressed in terms of kT) being of the order of 1 eV while electrons will reach up to about 100 keV.

This hot environment, characterized by an electron density of 10^{12}-10^{14} cm^{-3} and a radioactive ion density of about 10^{11} cm^{-3} will be used to investigate the predicted evolution of the β decay rate as a function of the plasma parameters. To reach this task the plasma will be maintained in a dynamic equilibrium by equalizing input fluxes to the losses from the magnetic confinement. When the plasma operates in magneto-hydrodynamic equilibrium, the dynamical equilibrium can be reached under proper tuning conditions of the magnetic field profile, background pressure, and radiofrequency (RF) power. The β decay events will be tagged detecting the γ-rays emitted by the daughter nuclei, still confined in the plasma, using an array of high efficiency High Purity Germanium (HPGe) detectors [10]. The in-plasma measured radioactivity will be directly correlated to the plasma thermodynamical properties (density and temperature) which will be monitored on-line using a multi-diagnostics setup.

The detection of gamma-rays emitted in the decay process of the daughter nuclei (populated after the beta decay) is a rather complex issue to achieve when a plasma trap is used. In fact, the trap poses many mechanical constraints related to the presence of the magnetic system and the cryostat which affect the positioning of the detectors. Detectors have to be placed in regions free of coils to be able to detect the gamma-rays without any shielding effect due to the presence of materials and, at the same time in regions of low magnetic field to avoid charge collection effect which could limit their performances.

The HPGe detectors will work in rather harsh conditions due to the X-rays' and γ-rays' background of about 50 kHz (in each detector) produced inside the plasma trap due to the electron bremsstrahlung. Dedicated electronics able to run at such a high rate without any significant worsening in energy resolution and efficiency have been already developed by the GAMMA collaboration, which will lend 16 HPGe detectors of the GALILEO Array [11] for the first experimental run of PANDORA.

Simulations focused on optimizing the detection efficiency led to the design of an array made of 14 HPGe detectors be placed around the trap, 12 radially and 2 axially (see fig 1b) [12]. Detectors will measure the γ-rays emitted the plasma through conical lines of sight made in the cryostat and the external yoke (see fig.1a) which pass through the center of the magnetic system. The total photopeak detection efficiency of the array reaches a maximum value of 0.17% at about 200 keV and then smoothly decrease down to 0.09% at about 1.5 MeV keeping the same value up to 2 MeV. Although the array efficiency seems to be rather low, such a value is compensated by the large number of decaying ions in the trap which allows to detect the expected variation of the radionuclide half-lives as shown by the GEANT4 simulations in a time ranging from a couple of weeks for the simplest cases to about 3 months for the longer-living isotopes, as a function of the ionization degree.

A crucial role in the project is played by the plasma multi-diagnostics which will work synergically with the HPGe detector array to on-line monitor main plasma parameters. The ionization states and the charge distributions are, in fact, determined by the plasma temperature, once fixed the density and the confinement time, and therefore their knowledge is fundamental to relate the plasma thermodynamical properties to the measured half-lives.

To deduce plasma properties one can exploit the radiation emitted by the plasma in a broad range of frequencies from radio to γ-rays. Its full characterization calls for the development of different tools to investigate all energetic domains and various analysis methods [13,14]. One important role is played by the shape of the electron energy spectrum which is characterized by the presence of three main components, a cold (E_e < 1 keV), a warm (1 keV < E_e < 30 keV) and a hot one (E_e > 30 keV). The warm one is playing a major role in the ionization process and can be monitored using a silicon drift detector allowing for volumetric soft X-ray spectroscopy which gives access to electron temperature and density. Similar information can be extracted for the hot and cold components using an HPGe detector and an optical spectrometer, respectively [15]. The multi-diagnostic system includes also an interfero-polarimeter [16] to measure the line-integrated total density and a RF probe coupled to a spectrum analyzer or a scope for time resolved spectroscopy [16] and for monitoring and mastering the plasma stability. A pin-hole camera will be used for high resolution spatially-resolved soft X-ray spectroscopy to investigate the plasma structure and the confinement dynamics in the range 2 – 20 keV [17].

In the first experimental campaign we will focus on the study of ^{176}Lu, ^{134}Cs and ^{94}Nb which have been selected among more than one hundred isotopes of potential interest for nuclear astrophysics. Theoretical predictions based on a model developed in [7] suggest that at the expected PANDORA electron temperatures (kT ~ 10 keV) their lifetimes may collapse by several orders of magnitudes. In particular, in the case of ^{94}Nb nucleus which provides the main production channel for ^{94}Mo through *s*-process, the predicted reduction at the PANDORA working condition is about 5 orders of magnitude. Simulation results of a real

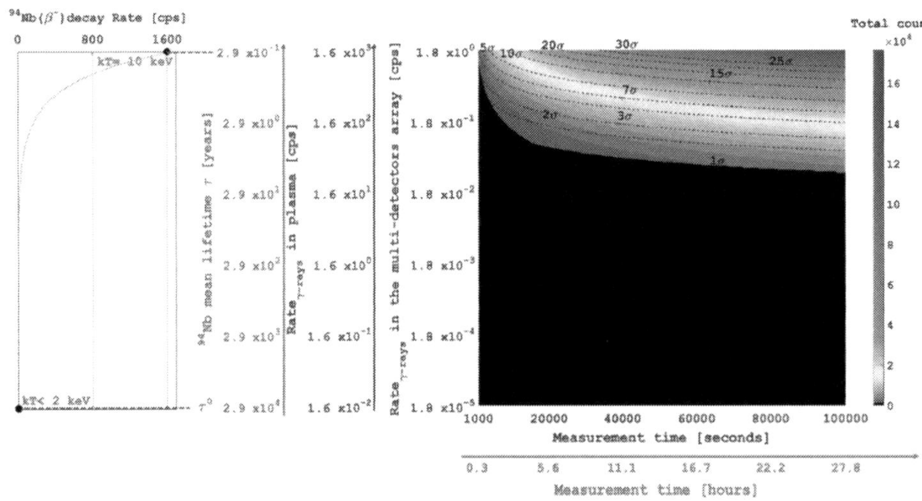

Fig.2. Measurability plot for ^{94}Nb isotope. Y-axis reports the expected variation of the lifetime (green axis); the corresponding expected rate in cps, considering the amount of isotope in the PANDORA plasma volume (blue axis); the corresponding rate of detected gammas in the multi-detectors array (black axis). The X-axis reports the measurement time, while the colors are the total counts with corresponding sigma-levels shown as dotted lines. The left part of the figure shows the decay rate in cps vs temperature.

experimental run assuming a concentration of 0.0001% of ^{94}Nb in a plasma volume of 1500 cm^3, including the efficiency of the HPGe detector array are shown in fig.2 [12]. The first (from the left) green vertical axis shows the range of lifetimes, expressed in years, that can be explored in the experiment, according to theory, starting from the lifetime of the neutral isotope. The predicted decay rate in the plasma is shown in the second, blue vertical axis. Finally, including the efficiency of the HPGe detector array the estimated counting rate of the detector array is shown in the third, black vertical axis. The x-axis shows the measurement time. Pseudo-colors show the total number of counts in the gamma peak (E_γ = 703 keV). To achieve a 3σ confidence level a run duration ranging from few hours to about one day, depending on the observed lifetime reduction, can be foreseen.

3 Conclusions

The PANDORA project is focused on the development of a new experimental setup to investigate the variation of the β-decay half-life of radioisotopes of astrophysical interest in a magnetized plasma. The setup, expected to come into operation in 2024, is based on three pillars, namely the compact plasma trap, the gamma detection array and the multi-diagnostic system which were described together with the first physics cases, ^{176}Lu, ^{134}Cs and ^{94}Nb. An important breakthrough in stellar evolution and nucleosynthesis models is foreseen if the predicted lifetime variations will be observed.

Acknowledgments

Authors are grateful to all the members of the PANDORA collaboration who contributed to the project development. The support of the INFN 3rd Nat. Comm. for its financial and scientific support to the PANDORA project is particularly acknowledged.

References

1. G.T. Emery, Annu. Rev. Nucl. Sci. **22**, 165 (1972)
2. E. Segrè, Phys. Rev. **71**, 274 (1947)
3. R. Daudel, Rev. Sci. **85**, 162 (1974)
4. M. Jung et al., Phys. Rev. Lett. **69**, 2164 (1992)
5. F. Bosch et al., Phys. Rev. Lett. **77**, 5190 (1996)
6. E. Fermi, Il Nuovo Cimento **11**, 1 (1934)
7. K. Takahashi and K. Yokoi, Nucl Phys. **A404**, 578 (1983)
8. D. Mascali et al., Universe **80**, 8 (2022)
9. G. Mauro et al., Front. Phys. Sec. Nucl. Phys. **10**, 931953 (2022)
10. A. Goasduff et al., Front. Phys. Sec. Nucl. Phys. **10**, 936081 (2022)
11. A. Goasduff et al., Nucl. Instrum. Methods A **1015**, 165753 (2021)
12. E. Naselli et al., Front. Phys. Sec. Nucl. Phys. **10**, 935728 (2022)
13. E. Naselli et al., J. Instrum. **14**, C10008 (2019)
14. S. Biri et al., J. Instrum. **13**, C11016 (2018)
15. A. Pidatella et al., Front. Phys. Sec. Nucl. Phys. **10**, 931744 (2022)
16. D. Mascali et al., Rev. Scient. Instr. **93**, 033302 (2022)
17. E. Naselli et al., J. Instrum. **17**, C01009 (2022)

RIB induced reactions: Studying astrophysical reactions with low-energy RI beam at CRIB

H. Yamaguchi[1,*], *S.* Hayakawa[1], *N.R.* Ma[1], *H.* Shimizu[1], *K.* Okawa[1], *Q.* Zhang[2,1], *L.* Yang[1,3], *D.* Kahl[1,4,5], *M.* La Cognata[6], *L.* Lamia[6,7,8], *K.* Abe[1], *O.* Beliuskina[1,9], *S.M.* Cha[10,11], *K.Y.* Chae[10], *S.* Cherubini[6,7], *P.* Figuera[6], *Z.* Ge[12,9], *M.* Gulino[6,13], *J.* Hu[14], *A.* Inoue[15], *N.* Iwasa[16], *A.* Kim[17,18], *D.* Kim[11,17], *G.* Kiss[12,19], *S.* Kubono[12,14], *M.* La Commara[20,21], *M.* Lattuada[6,7], *E.J.* Lee[10], *J.Y.* Moon[22], *S.* Palmerini[23,24], *C.* Parascandolo[21], *S.Y.* Park[17], *V. H.* Phong[12,25], *D.* Pierroutsakou[21], *R.G.* Pizzone[6], *G.G.* Rapisarda[6], *S.* Romano[6,7,8], *C.* Spitaleri[6,7], *X.D.* Tang[14], *O.* Trippella[23,24], *A.* Tumino[6,13], *N.T.* Zhang[14], *Y.H.* Lam[14,26], *A.* Heger[27], *A.M.* Jacobs[28,29], *S.W.* Xu[14], *S.B.* Ma[14], *L.H.* Ru[14], *E.Q.* Liu[14], *T.* Liu[14], *C.B.* Hamill[4], *A.* St J. Murphy[4], *J.* Su[30], *X.* Fang[31], *M.S.* Kwag[10], *N.N.* Duy[10], *N.K.* Uyen[10], *D.H.* Kim[10], *J.* Liang[32], *A.* Psaltis[32,33], *M.* Sferrazza[34], *Z.* Johnston[28,29], and *Y.Y.* Li[2]

[1] Center for Nuclear Study (CNS), University of Tokyo, Wako, Japan
[2] School of Nuclear Science and Technology, Lanzhou University, Lanzhou, China
[3] China Institute of Atomic Energy, Beijing, China
[4] School of Physics and Astronomy, the University of Edinburgh, Edinburgh, UK
[5] Extreme Light Infrastructure – Nuclear Physics, Horia Hulubei National Institute for R&D in Physics and Nuclear Engineering (IFIN-HH), Bucharest-Măgurele, Romania
[6] Istituto Nazionale di Fisica Nucleare - Laboratori Nazionali del Sud, Catania, Italy
[7] Department of Physics and Astronomy, University of Catania, Catania, Italy
[8] Centro Siciliano di Fisica Nucleare e Struttura della Materia, CSFNSM, Catania, Italy
[9] Department of Physics, University of Jyväskylä, Jyväskylä, Finland
[10] Department of Physics, Sungkyunkwan University, Suwon, Korea
[11] Center for Exotic Nuclear Studies, Institute for Basic Science, Daejeon, Korea
[12] The Institute of Physical and Chemical Research (RIKEN), Wako, Japan
[13] Faculty of Engineering and Architecture, Kore University of Enna, Enna, Italy
[14] Institute of Modern Physics, Chinese Academy of Sciences, Lanzhou, China
[15] Research Center for Nuclear Physics (RCNP), Osaka University, Osaka, Japan
[16] Department of Physics, Tohoku University, Sendai, Japan
[17] Department of Physics, Ewha Womans University, Seoul, Korea
[18] Korea University, Seoul, Korea
[19] Institute for Nuclear Research (Atomki), Debrecen, Hungary
[20] Department of Physics, University of Naples Federico II, Naples, Italy
[21] Istituto Nazionale di Fisica Nucleare - Section of Naples, Naples, Italy
[22] Rare Isotope Science Project, Institute for Basic Science, Daejeon, Korea
[23] Istituto Nazionale di Fisica Nucleare - Section of Perugia, Perugia, Italy
[24] Department of Physics and Geology, University of Perugia, Perugia, Italy
[25] Faculty of Physics, VNU University of Science, Hanoi, Vietnam
[26] School of Nuclear Science and Technology, University of Chinese Academy of Sciences, Beijing, China
[27] School of Physics and Astronomy, Monash University, Victoria, Australia
[28] The Joint Institute for Nuclear Astrophysics, Michigan State University, East Lansing, MI, USA

*e-mail: yamag@cns.s.u-tokyo.ac.jp

[29] Department of Physics and Astronomy, Michigan State University, East Lansing, MI, USA
[30] College of Nuclear Science and Technology, Beijing Normal University, Beijing, China
[31] Sino-French Institute of Nuclear Engineering and Technology, Sun Yat-Sen University, Zhuhai, China
[32] Department of Physics & Astronomy, McMaster University, Ontario, Canada
[33] Technische Universität Darmstadt, Darmstadt, Germany
[34] Département de Physique, Université Libre de Bruxelles, Bruxelles, Belgium

Abstract. Astrophysical reactions involving radioactive isotopes (RI) often play an important role in high-temperature stellar environments. The experimental studies on the reaction rates for those are still limited mainly due to the technical difficulties in producing high-quality RI beams. A direct measurement of those reactions would be still challenging in many cases, however, we can make a reliable evaluation of the reaction rates by an indirect method or by studying the resonance proterties. Here we introduce recent examples of experimental studies on such RI-involving astrophysical reactions, performed at Center for Nuclear Study, the University of Tokyo, using the low-energy RI beam separator CRIB. One is for the neutron-induced destruction reactions of ^{7}Be in the Big-Bang nucleosynthesis, and the other is the study on the ^{22}Mg(α, p) reaction relevant in X-ray bursts, which was performed with the resonant scattering method from the inverse reaction channel.

1 Introduction

Astrophysical reactions involving radioactive isotopes (RI) often play an important role in explosive stellar environments. Although the RI are seldom seen on the earth due to the finite lifetime, they do exist in stars, and contribute to the evolution and thermal dynamics of stellar objects. In this context, many experimental studies have been made on such RI-involving reactions, in spite of the technical difficulties. In a normal experimental condition, short-lived RI can only be used as the beam, not as the target. Then the reaction measurements could suffer from the limitation of the beam intensity, since the typical RI beam intensity is as small as 10^5 particles per second (pps) or less, while $> 10^{14}$ pps is available for light-ion beams. This great difference in the beam intensity is fundamental for the feasibility of the measurement.

Here we discuss possible approaches to study RI-involving reactions, introducing recent representative results from the low-energy RI beam facility CRIB [1–3] of the University of Tokyo. CRIB is an RI beam separator of Center for Nuclear Study (CNS), the University of Tokyo, located at the RIBF facility of RIKEN Nishina Center. CRIB can produce low-energy (< 10 MeV/u) RI beams by the in-flight technique with primary heavy-ion beams accelerated at the AVF cyclotron of RIKEN. A diagram to show the RI beams ever produced at CRIB is found in [3]. Most of the RI beams are produced via 2-body reactions such as (p, n), (d, p) and (^{3}He, n), taking place at an 8-cm-long gas target with a maximum pressure of 760 Torr.

The typical intensity of the RI beam at CRIB is 10^4–10^6 pps. The maximum intensity attained was 2-3 $\times$ 10^8 pps by the ^{7}Be beam produced with a cryogenic target system, in which the target gas can be cooled down to about 90 K [4]. One problem in producing such intense RI beams with a gas target is the density-reduction effect. A high-current beam can deposit heat to disperse the target gas around the beam track, resulting in a reduction of the local gas density along the beam. According to a previous work [5], a density reduction by 30 % was observed with a beam depositting heat of 60 mW/mm. We constructed a target system with a feature of forced circulation of the gas [4], by which we found a flow rate of 55 standard liters per minute (slm) was effective in eliminating the density reduction of the hydrogen gas, caused by a heat deposition of 65 mW/mm with the ^{7}Li primary beam. A high heat-proof target to further increase the RI beam intensity is under development. So far we have performed preliminary studies on the material dependence of the heat capacity, by changing the materials of the gas cell and its gas-sealing foils. Changing the sealing foils from Havar to molybdenum, the heat capacity was found to be doubled.

We introduce below two latest works at CRIB [6, 7], in which RI-involving astrophysical reactions were indirectly studied at astrophysical energies.

2 ^{7}Be+n reaction measurement with the Trojan horse method

The primordial production of light nuclides is well described with the standard model of Big Bang nucleosynthesis (BBN), however, the ^{7}Li/H abundance remains overestimated by a factor of 3–4, known as the cosmological lithium problem. We have studied the ^{7}Be$(n, p_0)^7$Li, ^{7}Be$(n, p_1)^7$Li*and ^{7}Be$(n, \alpha)^4$He reactions, which may affect the primordial ^{7}Li abundance [6]. To study the reactions of two unstable particles (^{7}Be and n), we applied the Trojan horse method (THM, see References in [6, 8]), which is an indirect method to study a 2-body reaction via a 3-body reaction measurement under a quasi-free kinematical condition. The ^{7}Be$(n, \alpha)^4$He reaction had been measured with the THM in a previous work [8], and our new work expanded the sensitivity also to the ^{7}Be$(n, p_0)^7$Li and ^{7}Be$(n, p_1)^7$Li* reactions, by achieving a sufficient energy resolution to separate the contributions of these two reactions.

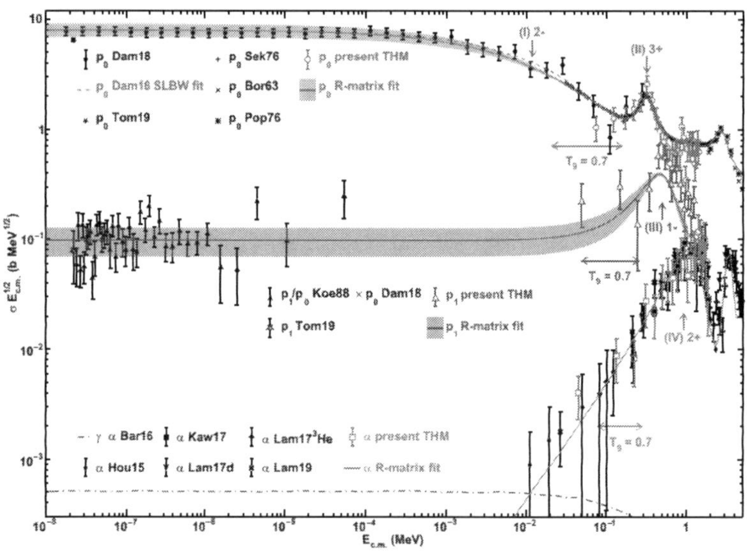

Figure 1. The cross sections multiplied by $\sqrt{E_{\rm cm}}$ of the ^{7}Be$(n, p_0)^7$Li, ^{7}Be$(n, p_1)^7$Li*and ^{7}Be$(n, \alpha)^4$He reactions in [6] are shown as open red circles, blue triangles, green squares, respectively. The previous experimental data are also shown for comparison. See [6] for details.

The cross sections multiplied by $\sqrt{E_{\rm cm}}$ obtained in this work are shown in Fig.1. The ^{7}Be$(n, p_0)^7$Li and ^{7}Be$(n, \alpha)^4$He rates were basically consistent with the previous works, but we assigned a smaller uncertainty on the ^{7}Be$(n, p_0)^7$Li reaction rate compared to the previous work at CERN n_TOF [9]. The ^{7}Be$(n, p_1)^7$Li* reaction was measured at the BBN energy for the first time in this work. With the simulation using `PRIMAT` code [10], we demonstrated the new reaction rate, including the ^{7}Be$(n, p_1)^7$Li* contribution, could reduce the primordial ^{7}Li abundance by $\sim$ 10–15%.

3 ^{25}Al+p elastic resonant scattering for the ^{22}Mg(α, p) reaction

(α, p) reactions often make significant contributions in explosive stellar environments, such as X-ray bursts. A typical situation is that the reaction has a positive Q-value and the rate is dominated by resonant reactions. In this case, the properties of the resonances just above the α-threshold would be the key information. Resonant scattering is one possible method to study resonances. However, in such cases the α-resonant elastic scattering from the entrance channel does not allow us to derive relevant resonance parameters, because those resonances are located at low center-of-mass energies, buried under the Rutherford scattering, and hardly seen. On the other hand, those resonances could be studied from the exit channel with the proton resonant scattering, since the proton threshold is located at a lower energy in this typical situation. We have carried out measurements with this idea [7, 11, 12], namely, the resonant scattering from the inverse reaction channel.

The latest study was for the ^{22}Mg(α, p) reaction, known to be one of the most relevant reactions in X-ray bursts, greatly affecting to the light curve. The reaction rate had not been precisely known, and the first direct measurement on this reaction has been performed only recently, but at an energy region higher than the Gamow energy [13]. Our measurement was performed at CRIB with an ^{25}Al RI beam at an intensity of 2×10^5 pps with essentially the same method as previous experiments [11, 14]. Fig. 2 shows the obtained excitation functions of the elastic and inelastic scatterings. We successfully observed resonances just above the α particle threshold, as indicated in the figure, and their parameters were deduced with an R-matrix analysis. The parameters were employed in a new X-ray burst model calculation, and we obtained a better agreement with the observed light curves of GS 1826–24 clocked burster and SAX J1808:4–3658 PRE burster compared to the previous works.

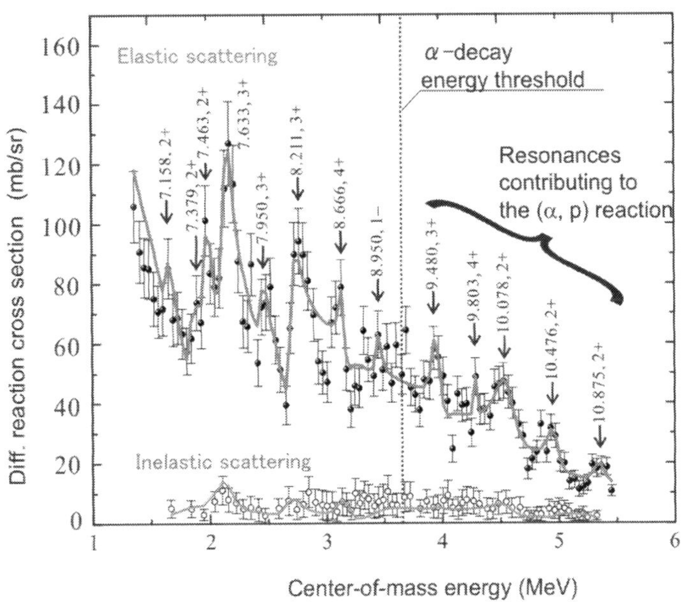

Figure 2. Excitation function of ^{25}Al+α scatterings with an R-matrix analysis. See [7] for details.

4 Summary

Astrophysical reactions involving RI are often difficult to study experimentally, mainly due to the limitation of the present RI beam technique. In the above two examples, we demonstrated that such astrophysical reactions can be studied even with a low-intensity RI beam, by employing an indirect method (THM) or studying resonance parameters.

The experiments were performed at RI Beam Factory operated by RIKEN Nishina Center and CNS, the University of Tokyo. This work was partly supported by JSPS KAKENHI (No. 18H01218, and 19K03883) from the Ministry of Education, Culture, Sports, Science and Technology (MEXT) of Japan.

References

[1] S. Kubono, Y. Yanagisawa, T. Teranishi, S. Kato, T. Kishida, S. Michimasa, Y. Ohshiro, S. Shimoura, K. Ue, S. Watanabe et al., Eur. Phys. J. **A13**, 217 (2002)

[2] Y. Yanagisawa, S. Kubono, T. Teranishi, K. Ue, S. Michimasa, M. Notani, J.J. He, Y. Ohshiro, S. Shimoura, S. Watanabe et al., Nucl. Instrum. Meth. Phys. Res., Sect. A **539**, 74 (2005)

[3] H. Yamaguchi, D. Kahl, S. Kubono, Nucl. Phys. News **30**, 21 (2020)

[4] H. Yamaguchi, Y. Wakabayashi, G. Amadio, S. Hayakawa, H. Fujikawa, S. Kubono, J. He, A. Kim, D. Binh, Nucl. Instrum. Meth. Phys. Res., Sect. A **589**, 150 (2008)

[5] J. Göerres, K.U. Kettner, H. Kräwinkel, C. Rolfs, Nucl. Instrum. Methods Phys. Res **177**, 295 (1980)

[6] S. Hayakawa, M.L. Cognata, L. Lamia, H. Yamaguchi, D. Kahl, K. Abe, H. Shimizu, L. Yang, O. Beliuskina, S.M. Cha et al., **915**, L13 (2021)

[7] J. Hu, H. Yamaguchi, Y.H. Lam, A. Heger, D. Kahl, A.M. Jacobs, Z. Johnston, S.W. Xu, N.T. Zhang, S.B. Ma et al., Phys. Rev. Lett. **127**, 172701 (2021)

[8] L. Lamia, M. Mazzocco, R.G. Pizzone et al., ApJ **879**, 23 (2019)

[9] L. Damone, M. Barbagallo, M. Mastromarco et al., PhRvL **121**, 042701 (2018)

[10] C. Pitrou, A. Coc, J.P. Uzan, E. Vangioni, Physics Reports **754**, 1 (2018)

[11] J.J. He, L.Y. Zhang, A. Parikh, S.W. Xu, H. Yamaguchi, D. Kahl, S. Kubono, J. Hu, P. Ma, S.Z. Chen et al., Phys. Rev. C **88**, 012801 (2013)

[12] J. Hu, J.J. He, A. Parikh, S.W. Xu, H. Yamaguchi, D. Kahl, P. Ma, J. Su, H.W. Wang, T. Nakao et al., Phys. Rev. C **90**, 025803 (2014)

[13] J.S. Randhawa, Y. Ayyad, W. Mittig, Z. Meisel, T. Ahn, S. Aguilar, H. Alvarez-Pol, D.W. Bardayan, D. Bazin, S. Beceiro-Novo et al., Phys. Rev. Lett. **125**, 202701 (2020)

[14] H. Yamaguchi, Y. Wakabayashi, S. Kubono, G. Amadio, H. Fujikawa, T. Teranishi, A. Saito, J. He, S. Nishimura, Y. Togano et al., Phys. Lett. B **672**, 230 (2009)

Overview of Numerical Simulations for Calculating In-Plasma β-Decay Rates in the Framework of PANDORA Project

Bharat Mishra[1,2,*] on behalf of the PANDORA collaboration

[1]Dipartimento di Fisica e Astronomia - Università degli studi di Catania, Italy
[2]Istituto Nazionale di Fisica Nucleare - Laboratori Nazionali del Sud, Catania, Italy

Abstract. β-decay rates are essential inputs in stellar nucleosynthesis models to explain observed nuclear abundances. While current models continue to use terrestrial values, experiments in storage rings indicate strong divergence between decay rates of neutral and ionised atoms [1], necessitating renewed investigations into stellar decay rates. The PANDORA project aims at measuring lifetimes of specific radio-isotopes trapped in an ECR plasma (which mimics astrophysical environments to some extent) and compare them with theoretical predictions [2], consequently verifying the models and allowing decay rate estimation for any isotope in the stellar interior. We present here a simulation scheme to characterise the space-resolved charge state and level population distribution of buffer and radio-isotope ions in ECR ion sources in order to calculate in-plasma decay rates. The algorithm is based on a Particle-in-Cell Monte Carlo (PIC-MC) routine that simultaneously models charge transport with collision-radiative processes. Preliminary results from the simulation are also shown, along with important takeaways for code-optimisation.

1 Introduction

Electron cyclotron resonance ion sources (ECRIS) are devices to produce ion beams using dense and energetic plasmas. Their operation relies on the dual concepts of magnetic confinement and resonance heating wherein electrons energised through resonance interaction with microwaves are trapped in a magnetic bottle (Fig. 1(a)) and collide with gas atoms to sequentially ionise them. Conventional ECR plasma chambers are filled with light "buffer" gases like H, He and Ar that either directly form the ions or "plasmise" injected heavy isotopes to generate radioactive ion beams. PANDORA (**P**lasmas for **A**strophysics, **N**uclear **D**ecay **O**bservations and **R**adiation for **A**rchaeometry) is an upcoming facility at INFN-LNS designed to exploit the high density ($n_e \sim 10^{11} - 10^{13}\,\mathrm{cm}^{-3}$) and temperature ($T_e \sim 0.01 - 100\,\mathrm{keV}$) of ECR plasmas to emulate astrophysical environments and study phenomena like plasma-induced modification of β-decay rates [3]. PANDORA intends to directly measure $T_{1/2}$ of radioactive nuclei like ^{176}Lu and ^{94}Nb diffused in the ECR plasma through secondary γ-tagging and compare the values with model predicted decay rates [2]. Once verified, the model can be adapted to stellar density and temperature conditions, and decay rates evaluated even for those isotopes which do not emit secondary γ photons.

*e-mail: mishra@lns.infn.it

Theoretical calculation of in-plasma $T_{1/2}$ requires inputs on ion charge state distibution (CSD) and level population distribution (LPD) which are difficult to evaluate in ECRIS because of system non-homogeneity and anisotropy. Additionally, many different physical processes affect the CSD and LPD, necessitating self-consistent evaluation of plasma properties and reaction rates. The main task then is to develop suitable models to extract 3D space-resolved CSD/LPD of buffer ions and radioisotopes, taking into account their transport in the plasma as well as reactions relevant to each. An overview of such an algorithm is reported in the following sections, whose details can be found in Ref. [5]. The algorithm is based on, and intended to be an extension of, an already existing steady state PIC code for modelling ECR plasma electron dynamics. The code is written in MATLAB© to take advantage of implicit matrix vectorisation offered by the language, and while it is mostly targeted towards use for PANDORA, its generic nature allows for expansion into fundamental ECRIS research as well.

2 Algorithm for Buffer Ions

The temporal evolution of population density in each charge state i is described by the standard ECRIS balance equation [4]

$$\frac{\mathrm{d}n_i}{\mathrm{d}t} = n_{i-1}n_e\gamma_{i-1,i} - n_in_e\gamma_{i,i+1} + n_{i+1}n_0E_{i+1,i} - n_in_0E_{i,i-1} - \frac{n_i}{\tau_i} \tag{1}$$

where n_i is the ion number density, n_e is the electron density, γ represents the collisional ionisation rate, n_0 is the neutral gas density, E is the single-electron charge exchange rate and τ_i is the characteristic confinement time of the charge state. In steady-state equilibrium, the time derivative of n_i vanishes, and all gain/loss processes in the RHS of Eq. 1 can be calculated independently to reconstruct the distribution of n_i. The simulation algorithm attempts to solve Eq. 1 using a PIC-MC code run sequentially for each i - N representative macroparticles are sampled according to $n_e\gamma_{i-1,i}$ and moved in a collisional plasma containing electromagnetic fields (Fig. 1(a)) according to Lorentz and Langevin equation. During their transport, $n_e\gamma_{i,i+1}$ and $n_0E_{i,i-1}$ are calculated and the trajectories of remaining particles mapped to accumulation maps. The accumulation maps of all charge states simulated until i are converted into density maps using charge neutrality to generate the CSD which converges to the true steady-state value as more charge states are simulated. By simultaneously modelling particle transport and atomic processes and tracking transfer between charge states, the algorithm captures the essential components of ECR ion dynamics and allows accurate reconstruction of density maps and ion CSD. The simulation scheme rests on 3 essential modules modelling, respectively, charge particle transport, MC-based interaction sampling and density scaling using transfer coefficients. Fig. 1(b) shows first results from the simulation in the form of 3D distribution of mean charge state - higher charges tend to accumulate near the axis, and all ions leave traces along the magnetic field lines on comparing with the field lines of Fig. 1(a).

3 Algorithm for Radioisotopes and Decay Rate

The PIC-MC scheme described in Sec. 2 can be identically applied to the radioisotope ions in order to calculate their 3D space-resolved CSD and LPD. The corresponding balance equation to be solved is modified to account for the disproportional concentration of the two ion species and a second set of equations added to model the distribution among levels i' *within*

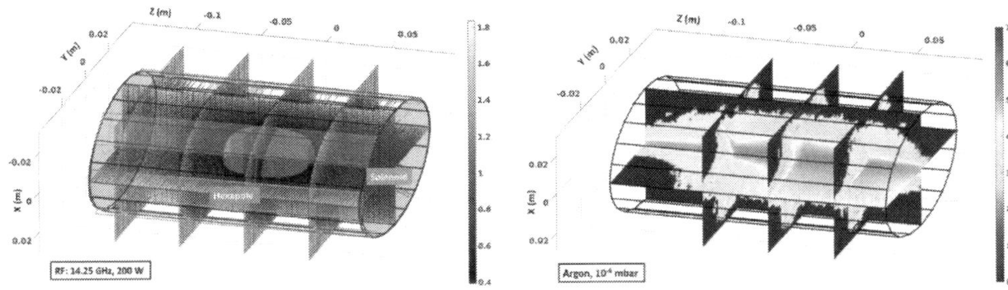

(a) (b)

Figure 1: (a) 3D magnetostatic field profile of typical ECRIS with resonance surface (green shell) and associated field streamlines (red curves) generated using solenoid (blue) and hexapoles (yellow) and (b) 3D map of mean charge of Ar plasma when simulated till 8^+. Operating conditions like microwave frequency, power and gas pressure are mentioned in insets.

a charge state.

$$\frac{\mathrm{d}n_i^R}{\mathrm{d}t} = \left[n_{i-1}^R n_e \gamma_{i-1,i} - n_i^R n_e \gamma_{i,i+1}\right] + \left[n_{i+1}^R \sum_{j=0}^{N-1} n_j E_{i+1,i} + n_{i-1}^R \sum_{j=1}^{N} n_j E_{i-1,i}\right] - \left[n_i^R \sum_{j=0}^{N-1} n_j E_{i,i-1} + n_i^R \sum_{j=1}^{N} n_j E_{i,i+1}\right] - \frac{n_i^R}{\tau_i^R} \quad (2)$$

$$\frac{\mathrm{d}n_{ii'}^R}{\mathrm{d}t} = \left[\sum_{ij'<ii'} n_{ij'}^R n_e C_{ij'ii'} - n_{ii'}^R \sum_{ij'>ii'} n_e C_{ii'ij'}\right] + \left[\sum_{ij'>ii'} n_{ij'}^R A_{ij'ii'} - n_{ii'}^R \sum_{ij'<ii'} A_{ii'ij'}\right] \quad (3)$$

Here n_i^R represents the isotope ion density in charge state i while $n_{ii'}^R$ is the same for level i', n_j is the buffer ion density distribution at the end of Sec. 2, and C and A respectively stand for collisional excitation and spontaneous emission rates. Eq. 2 differs from Eq. 1 in that self-collisions *between* isotope atoms are neglected and charge exchange is evaluated through collisions with all charge states n_j of the buffer ion. Eq. 3 is the added balance equation aiming to implement the so-called "coronal" model to describe level population distribution using electron collision excitation and spontaneous emission.

Once $n_{ii'}^R$ is known, it can be converted to its equivalent probability distribution $p_{ii'}^R$ and used to calculate the position-dependent decay rate λ_{tot} as [2]

$$\lambda_{tot} = \left[\frac{\ln 2}{f_{IF(m)} t_{1/2}}\right] \sum f_{IF(m)}^{*} \quad (4)$$

$$f_{IF(m)}^{*} = \begin{cases} \sum p_{ii'} \int_{1}^{W_{max}(ii')} (W^2-1)^{1/2} W (W_{max}(ii') - W)^2 F_0 S_{(m)}(ii') f_d(ii') \mathrm{d}W & \text{continuum decay} \\ \sum p_{ii'} \int_{W_{min}(ii')}^{\infty} (W^2-1)^{1/2} W (Q(ii')/m_e c^2)^2 F_0 S_{(m)}(ii') f_c(ii') \mathrm{d}W & \text{continuum capture} \\ \sum p_{ii'} \sum_{x(ii')} \sigma_x \frac{\pi}{2} [g_x \text{ or } f_x]^2 (Q(ii')/m_e c^2)^2 S_{(m)x(ii')} & \text{bound decay/capture} \end{cases} \quad (5)$$

where $f_{IF(m)}$ is the lepton phase volume and $t_{1/2}$ is the decay half-life measured on Earth, $f_{IF(m)}^{*}$ is the in-plasma lepton phase volume with m denoting type of transition, W, W_{max} and W_{min} are energy variables connected to the Q-value $Q(ii')$, F_0 is the Fermi function,

$S_m(ii')$ is the shape factor, f_d, f_c are connected to the Fermi-Dirac distribution function, m_e is electron rest mass, $x(ii')$ denotes electron orbital configuration with charge state i and level i', σ_x is occupancy probability and $f(x)$, $g(x)$ represent the larger of electron wavefunction evaluated at nuclear surface. The summation in Eq. 4 indicates the contribution of multiple decay channels to total rate λ_{tot}. Since the variation of λ_{tot} with individual levels i' is small and depends more strongly on averages over several levels, grouping can be done based on type of decay, reaction energetics and kinetics to simplify Eq. 3 and gain computational time. This may be particularly useful for for heavy isotopes which have complex electronic configurations.

4 Future Upgrades

The simulation algorithm is being continuously upgraded from both physical and computational points of view. One of the main improvements involves updating Eq. 1 to include a more complete exchange module, one that considers collisions between charge states and modifies n_0 dynamically. Other tasks foreseen are inclusion of electron capture, collisional de-excitation and implementation of radiation transport equation (RTE) in all balance equations. There are also efforts underway to upgrade from a particle-in-cell to cloud-in-cell approach to enhance numerical precision, as well as attempts to port the algorithm to GPUs.

Acknowledgements

The author is grateful for the financial support of PANDORA_Gr3 experiment under the 3rd Nat. Comm. of INFN.

References

[1] F. Bosch *et al*, Physical Review Letters **77**, 5190-5193 (1996).
[2] K. Takahashi and K. Yokoi, Nuclear Physics A **404**, 578-598 (1983).
[3] D. Mascali *et al*, Universe **8**, 1-16 (2022).
[4] G. Melin *et al*, Journal of Applied Physics **86**, 4772-4779 (1999).
[5] B. Mishra *et al*, Frontiers in Physics **10:932448**, 1-17 (2022).

Study of (α,n) reactions to determine the α-optical potential for γ-process

Dipali Basak[1,2,*] and *Chinmay* Basu[1]

[1]Saha Institute of Nuclear Physics, 1/AF, Bidhannagar, Kolkata-700064, India
[2]Homi Bhabha National Institute, Anushaktinagar, Mumbai, Maharashtra-400094, India

Abstract. The optical potential is calculated from elastic scattering angular distribution data at high energies. The potential derived from elastic scattering data does not satisfactorily describe the experimental result for low energy reactions. An alternative approach for improving the α-optical potential at low energy was proposed, based on the (α,n) reaction, which is solely dependent on the α- transmission coefficient, i.e. on the alpha optical potential. The new potential is a modified McFadden-Satchler (McF) potential with depth parameters (both real and imaginary) changed by an energy dependent Fermi function. The sensitivity of (α,n) cross-section with different width parameters is also studied.

1 Introduction

The neutron capture process produces nuclei heavier than iron in the stars. But 30-35 proton-rich heavy nuclei from ^{74}Se- ^{196}Hg (known as p-nuclei) are formed by the γ-process, instead of neutron capture. γ-process is a combination of (γ,α), (γ,n), and (γ,p) reactions on heavy nuclei in a high γ-flux scenario. The (γ,α) reaction rate is calculated from the inverse reaction using the principle of detailed balance. The Hauser Feshbach (HF) statistical model is used to determine the (α,γ) reaction rate from reaction cross-section data. α-optical model potential (α-OMP) is one of the primary input parameters for HF statistical model calculation.
There were various global α-OMPs available [1–3]. These potentials are obtained by fitting the α-elastic scattering angular distribution data and predicting reaction data at higher energies. However, it fails to explain the lower energy reaction data. Therefore, a modification in potential is required for low-energy astrophysical reactions. In this work, an energy-dependent α-OMP form was obtained by fitting the (α,n) reaction cross-section data of p-nuclei in the mass range A ~ 92 - 168.

2 Reaction Cross section: Statistical Model

In HF theory, alpha induced reaction cross section for a specific outgoing channel x (x being γ, n, p, α, etc.) is $\sigma_{\alpha x} = \sigma_\alpha$ P_x. Where σ_α is the compound nucleus formation cross section. Decay probability (P_x) of the outgoing channel x is given as $P_x = T_x / \sum_i T_i$, where T is the transmission cofficient. For neutron channel T_n is much higher than other particles (α,p, etc.), as there is no coulomb barrier for neutron. So, above the neutrons threshold $P_x \approx 1$ and (α,n) reaction cross section becomes $\sigma_{\alpha n} \sim \sigma_\alpha \sim \pi \lambda_\alpha^2 (2L+1) T_\alpha$. As T_α is calculated from entrance channel optical potential, (α,n) reaction is used to derive the α-OMP.

*e-mail: dipali.basak@saha.ac.in

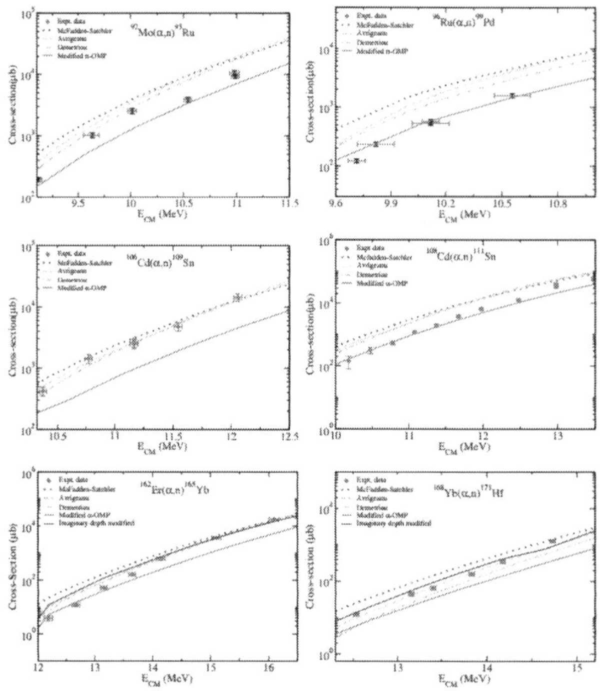

Figure 1. Expt. (α,n) reaction data [5–10] compared with the HF calculations using different α-OMP.

3 Results

The statistical model code TALYS in version 1.95 [4] is used for the HF calculations. Six even-even p-nuclei in the mass range A ~ 92-168 have been chosen. At first, HF calculations have been performed using well known global potentials and compared with the experimental results. The theoretical calculations suitably explained ^{106}Cd data but overestimated for other nuclei. A modified potential was introduced by the fitting of the (α,n) reaction data. This potential is basically the well known McFadden-Satchler global α-OMP [1] with its depth parameters (both real and imaginary) modified by an energy dependent fermi function f = 185 / [1 + $exp(0.9E_{cu} - E_{cm} / a_c)$], where E_{cu}, E_{cm} are in MeV and a_c = 10 MeV. E. Somorjai et al. [11] first used this modification, but only modifying the imaginary depth on ^{144}Sm. The HF calculations using different α-OMPs are shown in Fig. 1.

4 Discussion

The new modified α-OMP properly described the (α,n) reaction data mainly in the lower mass region. However, modifying the imaginary depth rather than the two depths provides better fitting in regions with larger mass. The cross-section sensitivities should be taken into consideration to properly interpret this different characteristic of the potential form. Effect on the cross-section (σ) values with the variation of particle or γ width (w) known as sensitivity (S), can be written as S = (F_σ - 1)/(F_w - 1). Where $F_\sigma = \sigma_{new} / \sigma_{old}$ and $F_w = w_{new} / w_{new}$. S = 1 implies that cross section changes with the same proportion as the width changes and S = 0 means no change occurs in cross section value. Sensitivity plots for (α,n) reaction are shown in Fig. 2. It was observed that (α,n) reactions for p-nuclei ^{92}Mo, ^{96}Ru and ^{108}Cd are mainly dependent on the α-width, other width contributions are negligible (< 25%). So, (α,n) cross

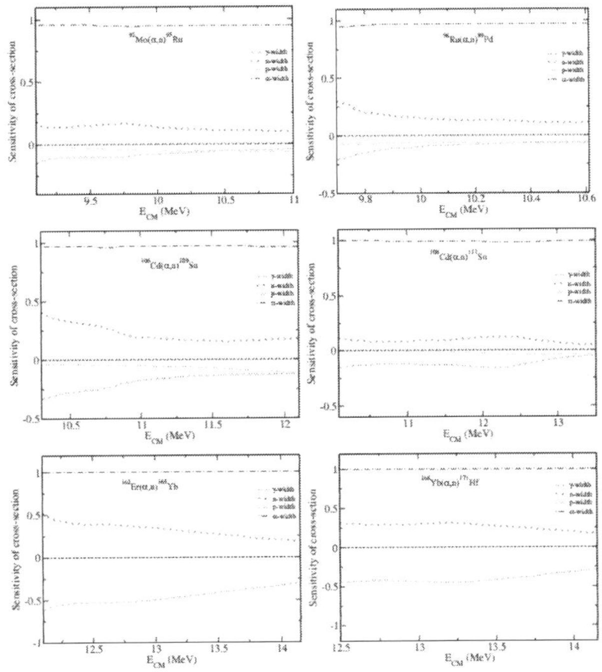

Figure 2. (α,n) cross section sensitivity with the variation of α-, p-, n-, γ-width [12].

section is directly dependent on alpha width i.e on α-OMP. The modified α-OMP adequately explains the results for these nuclei. But for other nuclei, besides the α-width (α,n) cross section is also sensitive to the neutron and γ width. So, it was difficult to extract α-OMP from (α,n) fitting for these higher mass nuclei. Surprisingly the results fit better for modifying only imaginary depth. The reason is that using this potential other width contribution may be eliminated. In Ref. [12], it is shown that this modified α-optical potential is explained (α, γ) reaction cross-section and reaction rate with proper level density model and γ-ray strength function.

References

[1] L. McFadden, G.R. Satchler, Nuclear Physics **84**, (1966) 177
[2] V. Avrigeanu et al., Physical Review C **90**,(2014) 044612
[3] P. Demetriou et al., Nuclear Physics A **707**, (2002) 253
[4] A.J. Koning, S. Hilaire, S. Goriely, *TALYS 1.95 A nuclear reaction program*, 2019
[5] W. Rapp et al., Physical Review C **78**, (2008) 025804
[6] W. Rapp et al., Physical Review C **66**, (2002) 015803
[7] Gy. Gyürky et al., Physical Review C **74**, (2006) 025805
[8] P. Scholz et al., Physical Letters B **761**, (2016) 247-252
[9] G.G.Kiss et al., Physics Letters B **735**, (2014) 40-44
[10] L.Netterdon et al., Nuclear Physics A **916**, (2013) 149
[11] E.Somorjai et al., Astron. Astrophys. **333**, (1998) 1112
[12] T. Rauscher, Ap. J. Suppl. **201**, (2012) 26
[13] Dipali Basak et al., Eur. Phys. J. A **58**, (2022) 150

Experimental study of $^{37}Cl(\alpha, n)^{40}K$ reaction in order to constrain the reaction rate of destruction of ^{40}K in stars

Nikolaos Dimitrakopoulos[1,5,*], *Georgios* Perdikakis[1,3,5]*Pelagia* Tsintari[1,5], *Carl R.* Brune[2,5], *Thomas N.* Massey[2,5], *Zach* Meisel[2,5], *Alexander* Voinov[2,5], *David C.* Ingram[2,5], *Panos* Gastis[4,5], *Yenuel* Jones-Alberty[2,5], *Shiv K.* Subedi[2,5], *Justin* Warren[2,5], *Kristyn H.* Brandenburg[2,5], *Nisha* Singh[2,5], and *Lauren P.* Ulbrich[1,5]

[1]Department of Physics, Central Michigan University, Mt. Pleasant, MI 48859, USA
[2]Department of Physics & Astronomy,Ohio University, Athens, OH 45701, USA
[3]Facility for Rare Isotope Beams, East Lansing, MI 48824, USA
[4]Los Alamos National Laboratory, Los Alamos, NM 87545, USA
[5]Joint Institute for Nuclear Astrophysics –Center for the Evolution of the Elements, East Lansing, MI 48824, USA

Abstract. ^{40}K is one of the main isotopes responsible for the radiogenic heating of the mantle in Earth-like exoplanets [1] and hence, plays a very important role in the internal geophysical dynamics of a planet. The abundance of ^{40}K in the mantle and the core of such planets is not always possible to be determined by astrophysical observations, although constraining the nuclear reaction rates of ^{40}K during stellar evolution can also lead to constraining the present amount of ^{40}K in these planets, which will improve our understanding on the habitability potential of Earth-like exoplanets. This study aims to constrain the ^{40}K(n,α)^{37}Cl reaction rate, one of the two major destruction paths of ^{40}K in stellar nucleosynthesis, by measuring the reverse reaction ^{37}Cl(α,n)^{40}K and applying the principle of detailed balance as we have done before for the ^{40}K (^{40}K(n,p)^{40}Ar reaction rate) [2]. During the first set of measurements we performed differential cross-section measurements of the ^{37}Cl(α,n$_1\gamma$)^{40}K, ^{37}Cl(α,n$_2\gamma$)^{40}K and ^{37}Cl(α,n$_3\gamma$)^{40}K reaction channels, for six different center of mass energies in the range between 5.1 and 5.4 MeV. The experiment took place at the Edwards Accelerator Laboratory of Ohio University. The gamma rays from the reaction channels mentioned above were detected by two LaBr3 scintillators. Using the swinger facility to change the angle of the beam-target system with respect to the detection system, we were able to take measurements for the differential cross-section at six different angles between 20° and 120° in the lab system.

*e-mail: dimt2n@cmich.edu

1 Introduction

Nuclei in the intermediate-mass region are of great interest for nuclear astrophysics [3], yet there is a great experimental deficiency. The unstable isotope of ^{40}K is one of the four main isotopes responsible for the radiogenic heating of the mantle in Earth-like exoplanets, in fact it dominates the planetary heat production about 5 Gyr after galaxy formation [1]. In addition, small quantities of the isotope may be present in the earth's core. The radiogenic heat keeps the mantle and the outer core (consisting mainly of molten iron and nickel) in a state of turbulent convection. The heat-induced motion of the core's outer layers generates the earth's magnetic field, which is essential for developing a habitable environment [4]. In addition, the heating of the mantle controls the formation of plate tectonics, the development of the volcanic activity, and the recycling of carbon on a planet, by controlling the amount of heat it will be extracted from the outer core to the mantle. The abundance of ^{40}K in a planet depends on the composition of the interstellar medium from which it was formed. Thus, nuclear reactions that determine the amount of ^{40}K during stellar evolution (both creation and destruction mechanisms) are not only crucial for understanding the fundamental mechanisms of potassium nucleosynthesis but may also play an essential role in understanding the habitability potential of earth-like exoplanets.

^{40}K is mainly created through three astrophysical processes: hydrostatic carbon shell burning, s-process, and explosive oxygen burning in type Ia supernovae [6, 7]. On the other hand, there are two main destruction paths for ^{40}K, the (n,p) and the (n,α) reactions. In this study, we aim to constrain the (n,α) destruction path, by measuring the time reverse reaction (^{37}Cl(α,n)^{40}K) and applying the principle of detailed balance.

2 Experimental setup

The experiment took place at the 4.5 MV tandem accelerator of the Edwards Acceleration Laboratory using the beam swinger capability of the facility [5]. A DC alpha particle beam was accelerated at energies between 5.65 and 5.95 MeV. The beam impinged into a thin ^{nat}KCl (1.1×10^{18} KCl/cm^2) target that was evaporated on a thin Au layer (1.34×10^{18} Au/cm^2). The composition and thickness of the target were determined via Rutherford Backscattering Spectroscopy, performed in the same facility after the experiment.

In this first measurement, we obtained the gamma rays originating from the de-excitation of the residual nucleus, hence we were able to perform differential cross section measurements for three reaction channels that corresponds to the three first excited states of ^{40}K, namely the ^{37}Cl(α,n$_1\gamma$)^{40}K, ^{37}Cl(α,n$_2\gamma$)^{40}K and ^{37}Cl(α,n$_3\gamma$)^{40}K. The three gamma rays that were measured, along with all the energeticaly allowed excited states of the residual nucleus are presented n Figure 1

For the gamma-ray detection, we placed two 2"x2" $LaBr_3$ detectors radially at 16 and 17 cm from the target, respectively. Both detectors were mounted on a stand to ensure a 52° relative angle between the two detectors in respect to the target. Finally, to determine the gamma-ray angular distribution for each of the reaction channels measured, the swinger magnet of the facility, which allows the beam axis to rotate around the target center, was utilized (Figure 2) enabling the measurement of gamma rays at 20°,45°, 60°,72°,95°, and 122°. Finally, to determine the efficiency of the $LaBr_3$ detectors, various gamma ray source measurements were performed after the experiment.

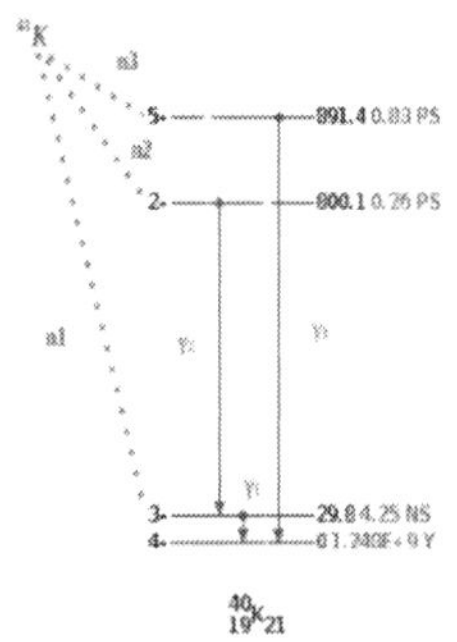

Figure 1. Energetically allowed excited states of the residual nucleus along with the three measured gamma rays .

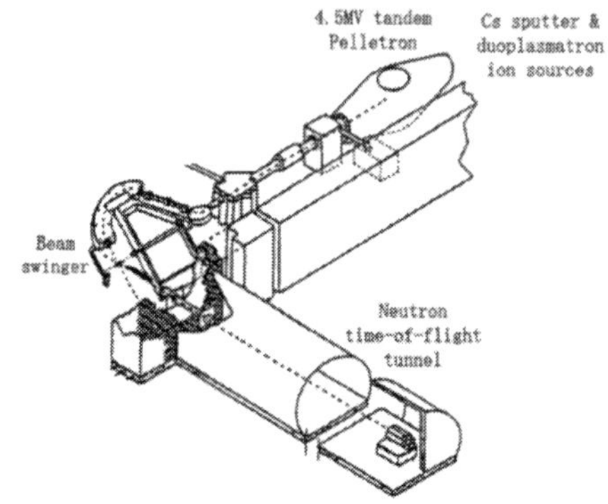

Figure 2. Beam swinger magnet diagram, Edwards Acceleration Laboratory.

3 Preliminary results and future perspectives

In this final section, some preliminary results are presented, along with the future plans for this ongoing study. In Figure 3 the differential cross sections for the reaction channels ^{37}Cl(α,n$_2\gamma$)^{40}K and ^{37}Cl(α,n$_3\gamma$)^{40}K at four different energies and six different angles are shown.Counting statistics is the primary source of the relative uncertainties, which range from 4 to 8 percent for each of these measurements. Furthermore, the gamma ray angular distributions for the two aforementioned reaction channels in two different alpha beam energies are presented in Figure 4. On the left plot, the angular distribution of the gamma rays originating from the de-excitation of the second energy level of ^{40}K at an alpha beam energy of 5.85 MeV is presented. The right plot shows the angular distribution for gamma rays from the third excited state when the target was irradiated with 5.95 MeV alpha particles. In both cases, the function used to fit the data was the first two even terms of Legendre polynomials, and in agreement with older measurements[8], the quadrupole radiation is dominant. The angular uncertainties are due to the diameter of the detectors and their distance from the target.

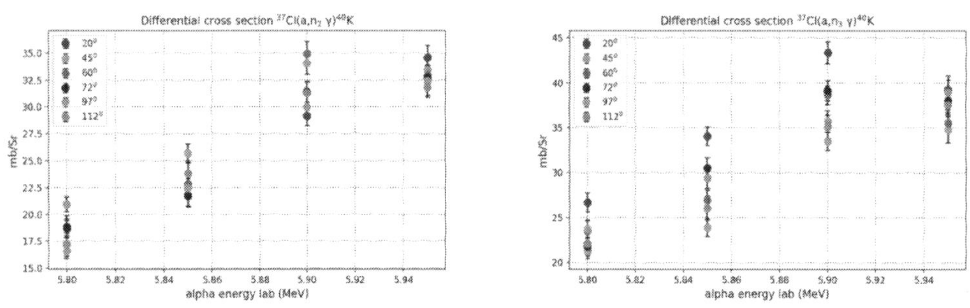

Figure 3. Differential cross sections for ^{37}Cl(α,n$_2\gamma$)^{40}K (left) and ^{37}Cl(α,n$_3\gamma$)^{40}K (right) reaction channels.

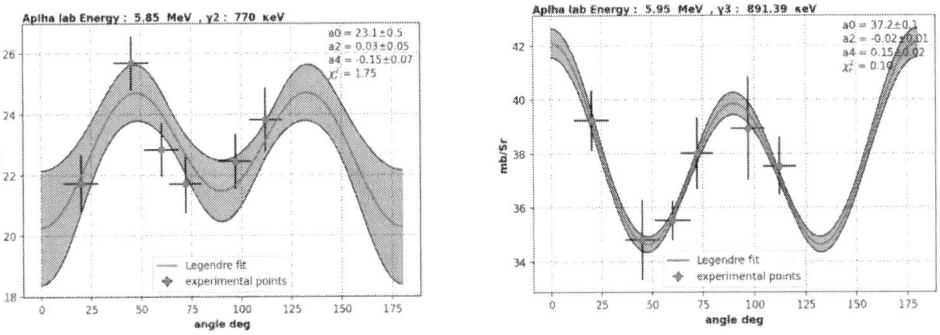

Figure 4. Gamma ray angular distributions from gamma rays originating from ^{37}Cl(α,n$_2\gamma$)^{40}K and ^{37}Cl(α,n$_3\gamma$)^{40}K reaction channels.

Our future perspectives for this work include performing an additional measurement. Along with the existing gamma ray detectors, an array of neutron detectors, namely organic liquid scintillators, will be used. Due to their excellent neutron gamma discrimination capabilities and their efficiency, we will be able to measure the ^{37}Cl(α,n$_0$)^{40}K reaction channel, that was impossible to measure with our current setup and ^{37}Cl(α,n$_1\gamma$)^{40}K. As for the latter, the internal self-activity of the $LaBr_3$ detectors imposed difficulties integrating the corresponding 29.8 keV gamma ray peak. Additionally, a code that reduces LaBr self activity, using a NaI as a reference detector, is currently under development[9]. Once we have measured the differential cross section for all the energetically available reaction channels of ^{37}Cl(α,n)^{40}K, we will be able to integrate them to obtain the total cross section and finally constrain the reaction rate of ^{37}Cl(α,n)^{40}K using the principle of detailed balance. Lastly, we will theoretically investigate the results and the alpha optical potential using Talys code [10], and the final results of this work will be presented in a future publication.

References

[1] Elizabeth A. Frank, Icarusl **243**, 274-286 (2014)

[2] Gastis P., Phys. Rev. C **101**, 16 (2020)

[3] Howard A. J, The Astrophysical Journal **188**, 131-139 (1974)

[4] U.R.Rao, Pramana-J Phy **15**, 33-43 (1980)

[5] Meisel Z., Conference on the Application of Accelerators in Research and Industry, Elsevier B.V (2016)

[6] Iliadis C. *Introduction to Nuclear Astrophysics*

[7] Truran J., The Astrophysical Journal **160**, 181 (1970)

[8] Wechsung R., Nuclear physics A **170**, 557-570 (1971)

[9] Ding Yuan, NIM-A **644**, 33-39 (2011)

[10] A.J. Koning,Proceedings of the International Conference on Nuclear Data for Science and Technology, 211-214 (2008)

The challenging direct measurement of the $65\,keV$ resonance strength of the $^{17}O(p,\gamma)^{18}F$ reaction at LUNA

Riccardo Maria Gesuè[1]* for the LUNA collaboration

[1]Università degli Studi di Napoli and INFN Division of Napoli, Via Cintia, 80126 Napoli (Italy)

Abstract. The $^{17}O(p,\gamma)^{18}F$ reaction plays a crucial role in AGB nucleosynthesis as well as in explosive hydrogen burning occurring in type Ia novae. At the temperatures of interest for the former scenario (20 MK < T < 80 MK) the main contribution to the astrophysical reaction rate comes from the poorly constrained E_R = 65 keV resonance. The strength of this resonance is presently determined only through indirect measurements, with an adopted value $\omega\gamma$ = (16 ± 3) peV.
A new high sensitivity setup was installed at LUNA, located at LNGS. The underground location of the LUNA 400kV accelerator guarantees a reduction of the cosmic ray background by several orders of magnitude. The residual background was further reduced installing a devoted shielding. On the other hand, to increase the efficiency, the 4π BGO detector was coupled with Al target chamber and holder. With more than 400 C accumulated on Ta_2O_5 targets, nominal ^{17}O enrichment of 90%, the LUNA collaboration has performed the first direct measurement of the 65 keV resonance strength.

1 Introduction

The $^{17}O(p,\gamma)^{18}F$ reaction takes part in the CNO cycle, active in the giant stars H-burning shell or in explosive burning scenarios (i.e. type Ia novae). The precise knowledge of the $^{17}O(p,\gamma)^{18}F$ nuclear reaction rate at typical temperatures of H-shell burning in AGB stars (20 MK $\leq T \leq$ 80 MK) is needed to address several astrophysical problems regarding the Oxygen isotopic ratio [1]. The post first dredge-up (FDU) Oxygen ratios predicted by theoretical models [2], often differ form observed surface abundances of low-mass giant stars ($M < 2M_\odot$) at the tip of the RGB on the Hertzsprung-Russell diagram. Moreover, in a group of pre-solar grains, are observed a moderate ^{18}O depletion and a high ^{17}O enrichment that cannot be retrieved by models. In this context the $^{17}O(p,\gamma)^{18}F$ reaction rate plays a key role, directly affecting the ^{17}O depletion and ^{18}O production (^{18}F decays with a $T_{\frac{1}{2}} = 109.77'$ in ^{18}O).

The past $^{17}O(p,\gamma)^{18}F$ reaction study performed at Laboratory for Underground Nuclear Astrophysics (LUNA) was mainly focused at Gamow energies for classical novae (160 keV $\leq E_{cm} \leq$ 370 keV), using both the prompt γ-ray detection and the activation method, leding to a precise determination of the E_R = 183 keV resonance strength [3] (resonance energies are expressed in the center of mass). However, the main contribution to the astrophysical reaction rate at temperatures of interest for the AGB star H-burning shell comes from the E_R = 65 keV

*e-mail: riccardo.gesue@na.infn.it

resonance. An accurate measurement of the resonance strength can improve the reaction rate determination and will help to constrain the RGB and AGB models. The strength of the $E_R = 65$ keV resonance is presently determined only through indirect measurements [4–6] providing the Γ_γ and Γ_α, while the Γ_p is derived from the $\omega\gamma_{(\alpha,\gamma)}$ [7]. The adopted value is $\omega\gamma = (16 \pm 3)$ peV.

Since the expected rate is as low as one reaction per Coulomb [8], a direct measurement of the $E_R = 65$ keV resonance strength required both a high sensitivity setup and a devoted technique for beam-induced background (BIB) suppression. In following sections the setup, the analysis and the preliminary results found will be described.

2 Experimental setup

The measurement was performed at LUNA, located in the deep underground facility of Laboratory Nazionali del Gran Sasso (LNGS). Thanks to the 1400 m overburden of rock, here the muon cosmic ray background is reduced by six orders of magnitude with respect to the surface [9]. Such a low cosmic-ray background in turn allows the deployment of thick passive shielding to absorb the natural radioactivity signals. In particular, a 15 cm thick lead shielding used for γ absorption was surrounded with a 5 cm borated (5%) polyethylene for n absorption: the polyethylene reduced the detected background of a factor 4 in the region of interest (ROI) of our measurement (5.2 - 6.2 MeV), with respect to using only lead. The LUNA 400kV accelerator [10] was able to provide a high stability 200 μA current on target at projectile energy $E_p = 80$ keV. The high efficiency (74% at 661 keV) Bismuth-Germanium-Oxide (BGO) detector was installed around the reaction chamber, covering a 4π angle around the target. The detector is made of six optically independent crystals, which coupled with a DAQ that records the time stamp of each reading allows both single crystal spectra and the construction of the add-back spectrum, namely by adding coincident events in the individual crystals. In order to minimize the γ-ray absorption, Aluminium was used as construction material for the scattering chamber and the target holder. A 3D model of the detection setup is shown in Fig.1.

The measurement was performed using Ta_2O_5 solid targets produced by anodization of tantalum backings in 90% ^{17}O enriched water. Targets were doped with 5% ^{18}O, thus allowing the periodical scan of the $E_R = 143$ keV $^{18}O(p,\gamma)^{19}F$ resonance to monitor target degradation and thickness.

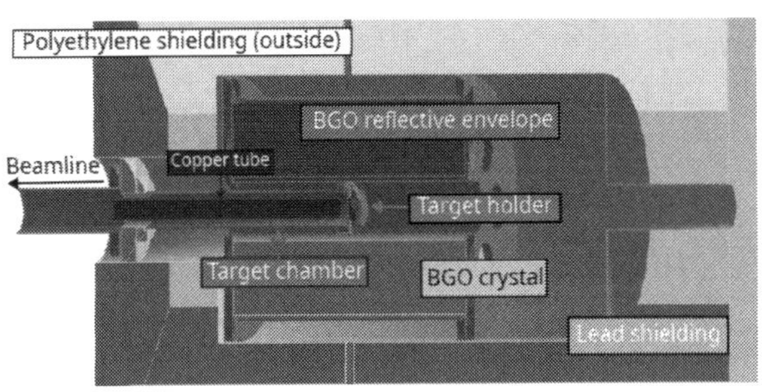

Figure 1. Section of the 3D model of the detector setup used in the Geant4 simulation.

3 Analysis and preliminary results

During five measurement campaigns, about 400 C were cumulated on ^{17}O targets. An accurate Montecarlo simulation of the setup was crucial for the analysis of the acquired data. The simulation was developed using the Geant4 toolkit [11] and it was optimized on the well-known spectra of ^{60}Co and ^{137}Cs sources, and ^{14}N$(p,\gamma)^{15}$O $E_R = 270$ keV resonance. See Fig.2 for a comparison between simulation and measurement of ^{14}N$(p,\gamma)^{15}$O. The average residuals between simulated and measured spectra after normalisation was below 3% in all three cases. This value also represents the precision at which the detection efficiency can be determined from the simulations.

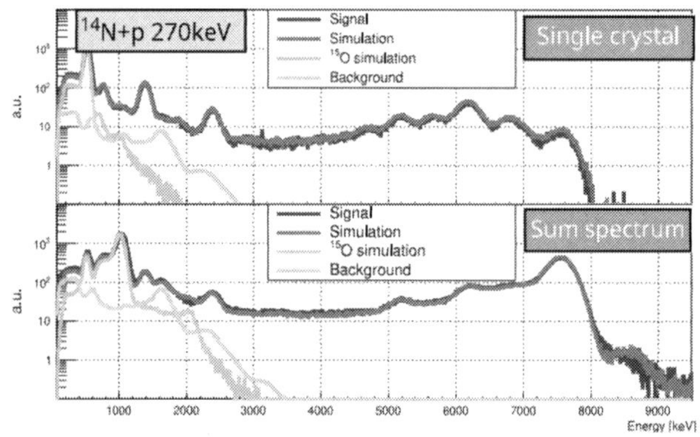

Figure 2. Comparison of simulated (red) and measured (blue) spectra of ^{14}N$(p,\gamma)^{15}$O 270 keV resonance. In green and yellow are represented environmental background and ^{15}O decay spectra respectively, included in the simulation.

A low (3 p.p.m.) deuterium contamination on the Ta backing produced a single γ peak at the same energy of the ^{17}O$(p,\gamma)^{18}$F 65 keV resonance peak. In Fig.3 the BIB peak is clearly visible and due to a contamination in the Ta backing. In order to subtract this BIB an innovative technique was developed which made use of our knowledge of the $E = 5672$ keV de-excitation branching ratios and of the BGO detector segmentation. The technique iso-

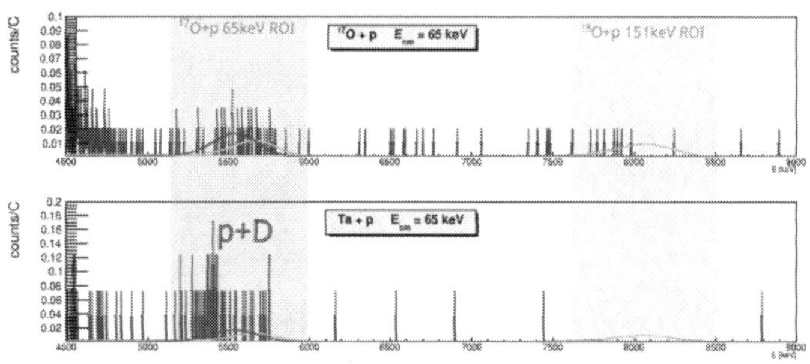

Figure 3. Spectra acquired shooting a $E_{lab} = 80$ keV proton beam on ^{17}O (top) targets and naked Ta backings (bottom). The $p + d$ peak in the bottom spectrum indicates a contamination in the Ta backings estimated at 3 p.p.m. level. The coloured peaks are shown to guide the eye.

lates multiplicity 2 and 3 transition γ-rays with the expected energies for the 5672 keV de-excitation chain from the multiplicity 1 BIB. This allows an almost complete background subtraction while losing only a small amount of resonance γ, since the probability of ground state transition (multiplicity 1) is 6%. The analysis lead to a preliminary $\omega\gamma$ determination of about double the value of the indirect estimates.

At the time of writing this proceeding, the last acquired data is being analysed and an in-depth evaluation of the uncertainties is being performed. A technical paper on the detector, setup and analysis technique is ongoing, while the final results will be published in a separate paper.

References

[1] A. Boeltzig, C.G. Bruno, F. Cavanna, S. Cristallo, T. Davinson, R. Depalo, R.J. deBoer, A. Di Leva, F. Ferraro, G. Imbriani et al., Eur. Phys. J. A **52**, 75 (2016)

[2] C. Abia, S. Palmerini, M. Busso, S. Cristallo, A&A **548**, A55 (2012)

[3] A. Di Leva, D.A. Scott, A. Caciolli, A. Formicola, F. Strieder, M. Aliotta, M. Anders, D. Bemmerer, C. Broggini, P. Corvisiero et al., Phys. Rev. C **89**, 015803 (2014)

[4] H.B. Mak, G. Ewan, H. Evans, J. MacArthur, W. McLatchie, R. Azuma, Nuclear Physics A **343**, 79 (1980)

[5] V. Landre, P. Aguer, G. Bogaert, A. Lefebvre, J.P. Thibaud, S. Fortier, J.M. Maison, J. Vernotte, Phys. Rev. C **40**, 1972 (1989)

[6] J.C. Blackmon, A.E. Champagne, M.A. Hofstee, M.S. Smith, R.G. Downing, G.P. Lamaze, Phys. Rev. Lett. **74**, 2642 (1995)

[7] M.Q. Buckner, C. Iliadis, K.J. Kelly, L.N. Downen, A.E. Champagne, J.M. Cesaratto, C. Howard, R. Longland, Phys. Rev. C **91**, 015812 (2015)

[8] G.F. Ciani, D. Piatti, R.M. Gesuè, for the LUNA collaboration, EPJ Web Conf. **260**, 11003 (2022)

[9] A. Best, A. Caciolli, Z. Fülöp, G. Gyürky, M. Laubenstein, E. Napolitani, V. Rigato, V. Roca, T. Szücs, Eur. Phys. J. A **52**, 72 (2016)

[10] A. Formicola, G. Imbriani, M. Junker, D. Bemmerer, R. Bonetti, C. Broggini, C. Casella, P. Corvisiero, H. Costantini, G. Gervino et al., Nuclear Instruments and Methods in Physics Research Section A: Accelerators, Spectrometers, Detectors and Associated Equipment **507**, 609 (2003)

[11] S. Agostinelli, J. Allison, K. Amako, J. Apostolakis, H. Araujo, P. Arce, M. Asai, D. Axen, S. Banerjee, G. Barrand et al., Nuclear Instruments and Methods in Physics Research Section A: Accelerators, Spectrometers, Detectors and Associated Equipment **506**, 250 (2003)

Experimental studies of the ^{46}Mn β^{+}-decay channel and spectroscopy of ^{46}Cr at LISE-GANIL

David Godos-Valencia[1,5,*], *Luis* Acosta[1,5,**], *Pauline* Ascher[2], *Bertram* Blank[2], *Jerome* Giovinazzo[3], *Francois* de Oliveira Santos[3], *Chloé* Fougères[4], and *Angel Miguel* Sánchez-Benítez[5]

[1]IF-UNAM, 04510 CDMX, Mexico
[2]LP2I Bordeaux, F-33175 Gradignan Cedex, France
[3]GANIL, 14076 Caen, France
[4]ANL, 60439 Illinois, USA
[5]CEAFMC-UHU, 21007 Huelva, Spain

Abstract. In this work we present the preliminary results of analysing the ^{46}Mn β^{+} decay channel as a way to study the ^{45}V(p,γ)^{46}Cr reaction. ^{46}Mn was selected among other species in the cocktail beam delivered by the LISE fragment separator at GANIL (Caen, France) in order to study its beta decay and the excited states of its daughter nucleus ^{46}Cr. As part of the validation process we present the ^{46}Mn half-life, the proton and gamma emission peaks related to the ^{46}Mn decay and compare them with the results from previous works.

1 Introduction

Stars with initial mass greater than 8 solar masses ($M_{\odot}$) end their lives through a Core Collapse Supernova (CCSN) explosion. Besides, ^{44}Ti nucleosynthesis takes place in CCSN making this nucleus a good gamma astronomy tracer for SuperNova (SN) events due to the characteristic γ-rays emitted in its decay chain (67.9, 78.4, 1157 keV). Furthermore, the comparison between observations and models of the synthetized ^{44}Ti in CCSN gives important constrains to the models. In the latter, reaction networks are used for modelling nucleosynthesis occurring in the last stages of those stars with thermonuclear reaction rates as its inputs [1–3].

Unfortunately, a direct measurement of the cross section for many thermonuclear reactions is extremely difficult in current laboratories worldwide. Therefore, indirect methods can be used for this purpose, especially when the reaction rate is dominated by a narrow isolated resonance. In this context, beta-delayed proton emission is very useful with (p,γ) reactions involving low and medium mass proton-rich radioactive nuclei. That is a consequence of the fact that in those reactions narrow isolated resonances are likely to occur [2, 4].

It is believed that nucleosynthesis of ^{44}Ti in CCSN explosions is quite sensitive to the ^{45}V(p,γ)^{46}Cr reaction [5]. In addition, there is no information of this reaction as it has not been studied before. Therefore, by exploring this reaction it is possible to give new information for the ^{44}Ti production in CCSN models. In the present analysis we propose to use the

*e-mail: dgodosv@gmail.com
**e-mail: acosta@fisica.unam.mx

^{46}Mn beta-delayed proton emission as a way to study the inverse reaction. For that purpose we analyze the data of the "Isospin mixing in pf-shell proton emitters" experiment (Code: E666, spokesperson: Bertram Blank, LP2I - Bordeaux France) performed with the LISE fragment separator at GANIL laboratory (Caen, France) focusing on the analysis of the ^{46}Mn isotope. In section 2 we discuss the analysis procedure and in section 3 we present our preliminary results, used for the validation of the procedure, for the half-life, and the proton and gamma energy spectra related to the β^+ decay of the ^{46}Mn [6, 7].

2 Experimental setup and analysis

In the E666 experiment, a primary beam of $^{58}Ni^{26+}$ at 74.5 MeV/u collided with a 230.6 mg/cm^2 thick ^{nat}Ni target. With the help of LISE, the isotopes to be implanted in the silicon detectors were selected. The detector array in the experiment (see figure 1) was composed by 4 HPGe clovers, each one divided into 4 HPGe crystals (labeled A, B, C, and D), surrounding 3 Si detectors: a ΔE detector of 316 μm thickness labeled as 2D6, a Double-Sided Silicon Strip Detector (DSSSD) of 297 μm where the fragments are implanted, and a SiLi detector of 5 mm used as veto for particles from the ion beam. In the beam line, upstream the detector array, was placed another Si detector labeled as 1D6 of 300 μm. Each of the sides of the DSSSD is divided in 16 strips, 3 mm width each, thus allowing for defining squared pixels 3 mm $\times$ 3 mm in size.

All the 32 signals of the DSSSD, along with the signals of the 1D6, 2D6, and SiLi detectors plus the 16 signals of all the HPGe crystals were preamplified, amplified and digitalized by Mesytec GmbH MPR64 preamplifiers, Mesytec GmbH STM amplifiers, and GANIL VXI electronics. The DAQ was triggered by the 1D6 and DSSSD detectors, saving all the data of the system alongside the information of the activated triggers.

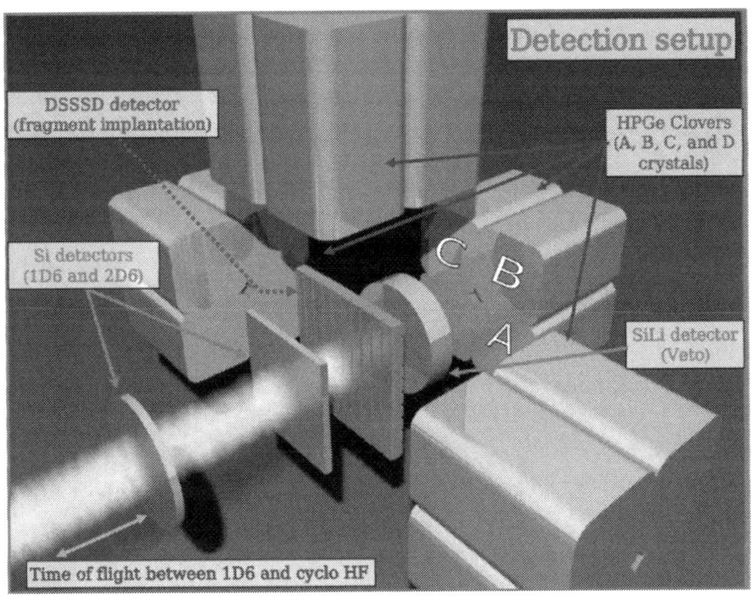

Figure 1. Detector array diagram. The HPGe clovers are depicted in blue, the Si and SiLi detectors in orange and the DSSSD in red. Image modified from [8].

2.1 Experimental analysis

We worked with a software developed at LP2I - Bordeaux complemented with a specific code, written using the CERN Root libraries, in order to complete the analysis. Firstly, we performed the calibrations of the DSSSD's X-Y strips and the clovers' HPGe crystals. A 3α source was used for the DSSSD and $^{56+60}$Co and ^{133}Ba+^{137}Cs sources were used for the clovers.

Secondly, the ion of interest was selected, ^{46}Mn, by means of an energy loss - time-of-flight 2D histogram (see figure 2). There we could easily identify, with the help of LISE simulations and previous works, the different ions implanted in the DSSSD. A graphical cut was used to make the selection of ^{46}Mn.

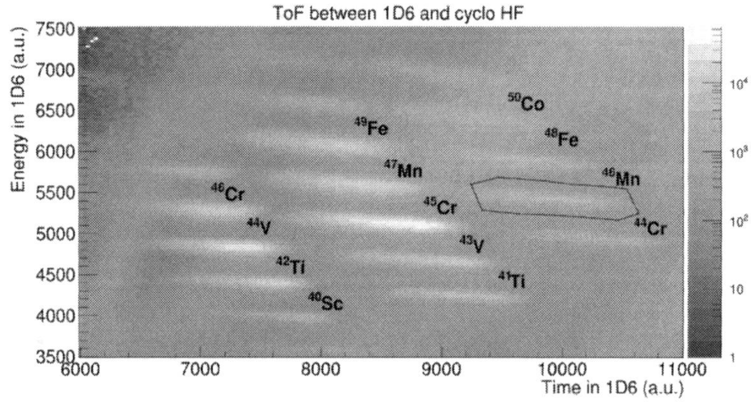

Figure 2. Energy loss in 1D6 detector vs time-of-flight 2D histogram used for the fragments identification of the nucleons implanted in the DSSSD detector.

Finally, time correlations between implantation and decay events were made. The correlations were established by opening a 1 s time window before and after a ^{46}Mn implantation event, and associating to it any decay event characterizing each correlation with a time difference ($\Delta t \equiv t_{\text{implantation}} - t_{\text{decay}}$). The negative Δt correlations were used to remove the background from the signal of the true correlation events. This allowed us to obtain the energy spectra for emitted protons and γ photons linked with the ^{46}Mn β^+-decay.

3 Discussion and results

By plotting the ^{46}Mn decay events with energy $E > 1500$ keV vs its Δt we obtained the decay curve, to which we calculated the half-life by fitting a 3 degrees of freedom exponential curve. We obtained $T_{1/2} = 37.5(4)$ ms with a $\chi^2/N = 1.01$, that is in agreement with $T_{1/2} = 36.2(4)$ ms obtained by [7] and $T_{1/2} = 34^{+4.5}_{-3.5}$ ms from [6]. It is important to note that we are only reporting the statistical error and it is pending to obtain the systematic one.

3.1 Proton and γ energy spectra

We also obtained the energy spectrum of protons emitted after the ^{46}Mn β^+-decay (see figure 3), and the emission spectrum for γ-rays (see figure 4). In both spectra it is possible to observe proton and γ peaks seen and listed by [7]; we are using for the peaks the same numeration as in his work. Nevertheless, it is interesting to note that in our results appear γ peak structures candidates at energies around 1, 1.08, and 1.17 MeV that are yet to be determined.

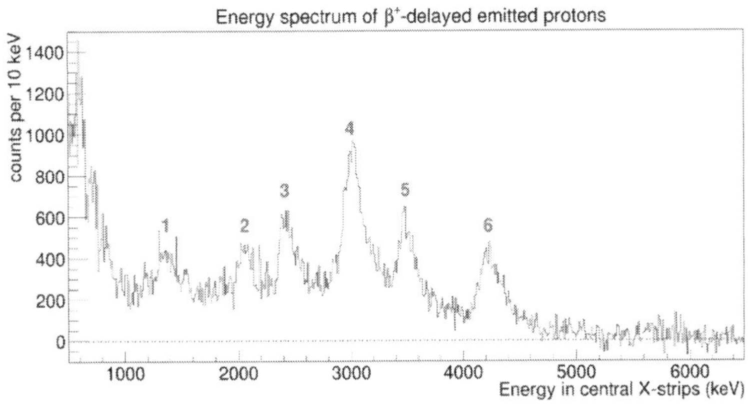

Figure 3. Sum-of-the-energy spectrum of the charged particles emitted after the ^{46}Mn β^+ decay for the central implantation pixels of the DSSSD. The red numbers list the proton peaks as they appear in the figure 22(b) of [7] Only the peaks 2, 4, 5, and 6 have an energy level assigned to them, while the peak 2 is associated with the decay of ^{45}Cr. The peaks 4-6 are assigned to the decay of the IAS to ^{45}V 4th, and 5th excited states, and to the ground state respectively.

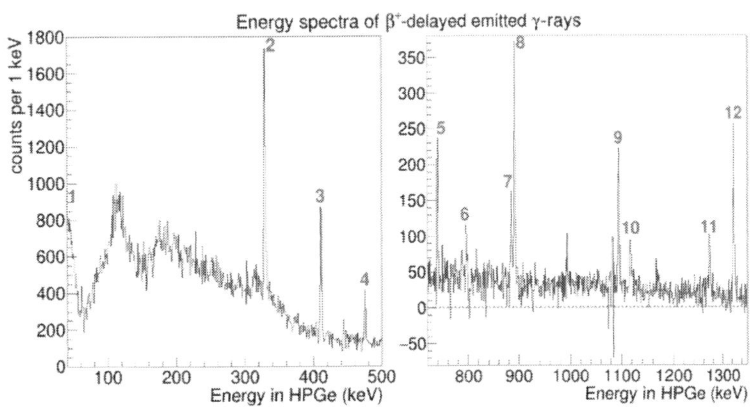

Figure 4. Sum-of-the-energy spectrum of the γ-rays emitted after the ^{46}Mn β^+ decay for all the HPGe crystals. The red numbers list the γ-rays peaks as they appear in the figure 23 of [7] The peaks 5, 6, and 10 have no energy levels assigned to them yet, while the peaks 8, and 9 are associated with ^{46}Cr ones. The other peaks are assigned to ^{45}V energy level transitions.

4 Conclusions

In this work we have obtained the ^{46}Mn β^+-decay time curve and calculated the half-life. The half-life is compatible with previous results from the literature. In addition, energy spectra for protons and γ-rays related to the decay were also obtained. Those spectra are also compatible with [6, 7] results, with greater resolution than either two publications. The above results obtained are part of the analysis validation for the ^{45}V(p,γ)^{46}Cr inverse reaction study. Weather we find at the end of the analysis any resonances contributing to the ^{45}V(p,γ)^{46}Cr reaction, according to [5] estimations, it would translate in a meaningful reduction of the ^{44}Ti production in CCSN.

The future steps for the analysis include the estimation of beta and γ-ray detection efficiencies of the silicon detectors and clovers respectively; and to find the branching ratios of the inverse reaction of $^{45}V(p,\gamma)^{46}Cr$. Besides, not all the peaks enumerated in figures 3 and 4 have a physical process assigned to them yet in the literature. The identification of such processes is still in progress.

Acknowledgements: This work is supported by DGAPA-UNAM IN107820 and CONACyT 314857 projects and by the Centro de Estudios Avanzados en Física, Matemáticas y Computación of the University of Huelva CEAFMC-UHU.

References

[1] A. Heger, et. al., ApJ **591**, 288-300 (2003)

[2] C. Illiadis, *Nuclear Physics of Stars* (Wiley-VCH, Germany, 2007) 672

[3] C. Giunti, and K. C. Wook, *Fundamentals of Neutrino Physics and Astrophysics* (Oxford University Press, Great Britain, 2007) 728

[4] L. Trache et. al., AIP Conference Proceedings **1409**, 67-70 (2011)

[5] L.-S. The et. al., ApJ **504**, 500-515 (1998)

[6] J. Giovinazzo et. al., Eur. Phys. J. A **10**, 73-84 (2001)

[7] C. Dossat et. al., Nuclear Physics A **792**, 18-86 (2007)

[8] C. Dossat, *Etude spectroscopique des noyaux riches en protons dans la region* $22 < Z < 28$ *et* $T_z < 3/2$ (Universitè Sciences et Technologies - Bordeaux I, France, 2004)

Astrophysical S factor and reaction rate of $^{92,94}Mo(p,\gamma)$ relevant to the p-process

Akash Hingu[1,*], *P.M.* Prajapati[2,3,**], *S.* Mukherjee[1,4,***], *R.G.* Pizzone[3], and *K.* Katovsky[5]

[1]Department of Physics, Faculty of Science, The M. S. University of Baroda, Vadodara-390002, *India*.
[2]Manipal Centre for Natural Sciences,Manipal Academy of Higher Education, Manipal-576014, *India*.
[3]Istituto Nazionale di Fisica Nucleare, Laboratori del Sud, Catania-95123, *Italy*.
[4]Department of Physics, Manipal Institute of Technology, Manipal-576014, *India*.
[5]Department of Electrical Power Engineering, Brno University of Technology, Brno, *Czech Republic*.

Abstract. Low energy proton capture cross sections on heavy isotopes are necessary for a better understanding of the astrophysical p-process. There are around 35 proton-rich stable isotopes between ^{74}Se and ^{196}Hg which are bypassed by the s- and r- processes. These are commonly referred as p-nuclei whose origin is still not completely understood. In the present study, proton capture reactions are studied on Mo isotopes at astrophysically relevant energies using nuclear modular code TALYS. Astrophysical S factor and reaction rates are also calculated inside a core-collapse supernova. The obtained results are compared with the literature data taken from EXFOR data library. In addition, the effect of different combinations of the nuclear input parameters entering the stellar reaction rate have been investigated.

1 Introduction

The proton capture process, often known as the "p-process", is the origin of several naturally occurring neutron-deficient isotopes of elements ranging from selenium to mercury. These nuclides are known as p-nuclei, and it is still unclear where they came from. The p isotope abundances seen in the solar system can not be replicated by the current p process [1] models. This can be partially explained by the ambiguous astrophysical conditions under which the process may take place. On the other hand, the failure might potentially be caused by deficiencies in the models of nuclear physics.

$^{92,94}Mo$ are such most abundant p-nuclei whose proton-capture reaction rate has been studied in the astrophysically relevant energy range. The reaction cross section is simulated using TALYS [2] nuclear modular code and compared with previous literature data available in EXFOR [3]. Astrophysical S factor is also calculated which shows greatly reduced energy dependence. Reaction rate for both the isotopes have been simulated and compared.

2 Calculations and Results

Theoretical calculations has been performed by the nuclear modular code TALYS-1.95 utilizing the Hauser-Feshabck (HF) statistical model. The code calculates cross section in the

*e-mail: hinguaakash@gmail.com
**e-mail: pareshkumar.p@manipal.edu
***e-mail: sk.mukherjee-phy@msubaroda.ac.in

energy range of 1 keV to 200 MeV taking into account all types of direct, pre-equilibrium, and compound mechanisms. The Hauser-Feshbach model is used to incorporate the compound reaction mechanism [4]. The pre equilibrium contribution is accounted by using the Kalbach exciton model [5]. The optical model parameters are calculated using Koning and Delaroche's suggested global potential [6]. Stellar reaction rate at different temperature has been calculated by using the TALYS code. This calculations is performed within the framework of statistical model because in stellar interior, nuclides do not only exist in ground state but also in different thermally excited states and a thermodynamic equilibrium holds locally to a very good approximation.

In general, the cross section strongly varies with the energy and declines rapidly to a very low value, making extrapolation to lower energies impractical. Therefore instead of the cross section, much less energy dependent quantity called astrophysical S factor is used for the extrapolation of reaction cross section which removes the fast coulomb dependence and is mainly sensitive to the nuclear effects [7]. The astrophysical S factor is defined by

$$S(E) = \frac{E}{e^{-2\pi\eta}} \sigma(E) \tag{1}$$

where η is sommerfeld parameter.

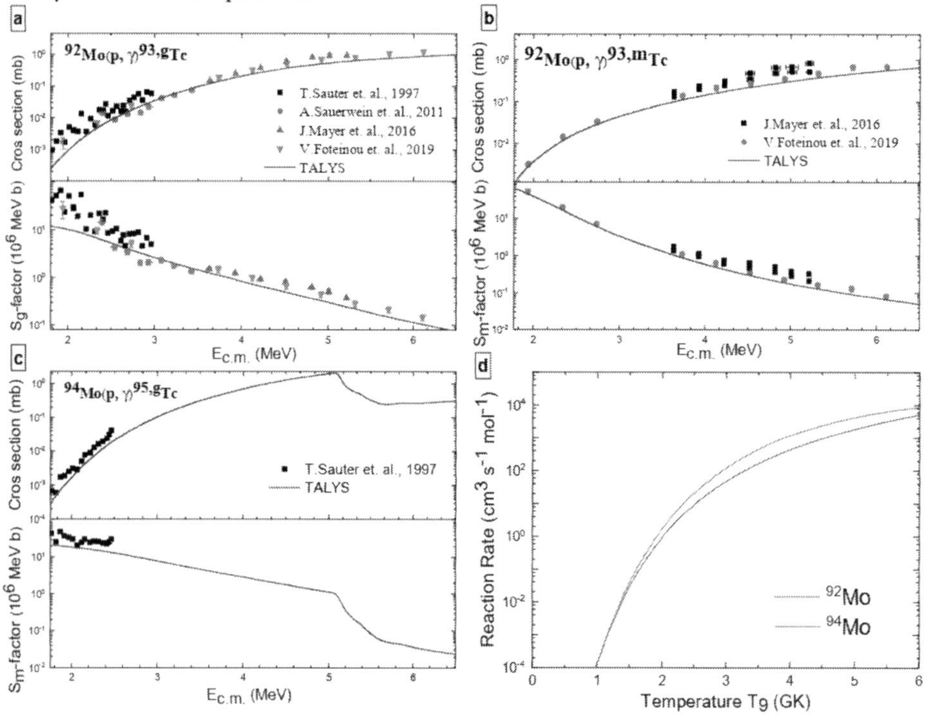

Figure 1. (Color Online) (a) Cross section and S factor for $^{92}Mo(p,\gamma)^{93,g}Tc$, (b) Cross section and S factor for $^{92}Mo(p,\gamma)^{93,m}Tc$, (c) Cross section and S factor for $^{94}Mo(p,\gamma)^{95,g}Tc$ and (d) Comparison of reaction rate for ^{92}Mo and ^{94}Mo.

The range of energies where the nuclear reactions occur in stars is given by the Gamow peak which is the overlap region of low energy tail of reaction cross section and the Maxwell Boltzmann distribution of the interaction particles. Gamow peak for $p +^{92} Mo$ at T = 2 GK is shown in the figure 2 - (a). The reaction rate was simulated for ^{92}Mo and ^{94}Mo using TALYS code. The comparison of reaction rate for ^{92}Mo and ^{94}Mo is shown in the figure 1 - (d).

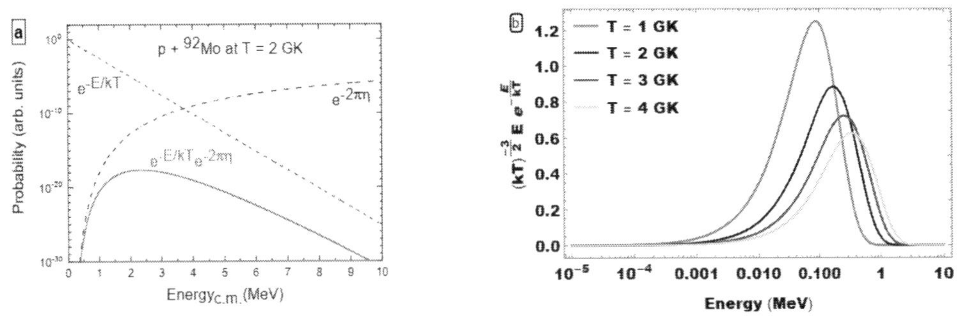

Figure 2. (Color Online) (a) Gamow window for p + ^{92}Mo at T = 2 GK and (b) Reaction rate equation parameters behaviour at different temperature.

Reaction rate equation has been studied and verified by using all the important parameters that is included in the equation. One such graph showing the variation of parameters at different temperature is shown in figure 2 - (b).

3 Summary and Conclusions

The cross section for the reactions $^{92}Mo(p,\gamma)^{93,g}Tc$, $^{92}Mo(p,\gamma)^{93,m}Tc$ and $^{94}Mo(p,\gamma)^{95,g}Tc$ has been simulated using TALYS and compared with the experimental results from the EXFOR data library, as shown in figure 1 - (a), (b) and (c) respectively. S factor with greatly reduced energy dependence is also shown with one to one correspondence to the cross section. Our theoretical predictions agree well with the experimental data. Moreover, there is no experimental cross section available for the $^{94}Mo(p,\gamma)^{95,g}Tc$ reaction in the EXFOR above 2.5 MeV as shown in figure 1 - (c), which reveals the scope of an experimental investigation. The reaction products are radioactive and the activation method [8] can be used to determine the cross section. Theoretical prediction or systematic approach [9] and indirect methods as well can be useful where reaction cross section is not directly possible to measure.

Acknowledgement

The author (AH) sincerely acknowledges the Department of Science and Technology (DST), Government of India for the INSPIRE Fellowship award (No. DST/INSPIRE Fellowship/2019/IF190924).

References

[1] M. Arnould, S. Goriely, Physics Reports **384**, 1 (2003)
[2] A. Koning, S. Hilaire, S. Goriely, *Talys user manual: A nuclear reaction program, nrg-1755* (2015)
[3] E.N.R. Data, CSISRS https://www-nds.iaea.org/exfor (2022)
[4] W. Hauser, H. Feshbach, Physical review **87**, 366 (1952)
[5] C. Kalbach, Physical Review C **33**, 818 (1986)
[6] A. Koning, J. Delaroche, Nuclear Physics A **713**, 231 (2003)
[7] C. Iliadis, *Nuclear physics of stars* (John Wiley & Sons, 2015)
[8] A. Hingu, B. Soni, S. Parashari, R. Makwana, P. Prajapati, V. Vashi, M. Mehta, R. Palit, S. Suryanarayana, B. Nayak et al., Radiation Physics and Chemistry **199**, 110270 (2022)
[9] A. Hingu, S. Parashari, S.K. Singh, B. Soni, S. Mukherjee, Radiation Physics and Chemistry **188**, 109634 (2021)

Nuclear level densities and γ-ray strength functions of ^{120,124}Sn and their application in astrophysics

Maria Markova[1,*], *Ann-Cecilie* Larsen[1], and *Frank Leonel* Bello Garrote[1]

[1]Department of Physics, University of Oslo, N-0316 Oslo, Norway

Abstract. Nuclear level densities (NLDs) and γ-ray strength functions (GSFs) of ^{120,124}Sn have been extracted with the Oslo method and additionally constrained with the novel Shape method. The GSFs were found to be in excellent agreement with the strengths from the inelastic relativistic proton scattering experiment. This comparison suggests the validity of the Brink-Axel hypothesis used as one of the key assumptions in astrophysical calculations. The extracted NLDs and GSFs were further used as experimental inputs to constrain the Maxwellian-averaged cross sections (MACS) for the radiative neutron capture process ^{119,123}Sn $(n,\gamma)^{120,124}$Sn using the the nuclear reaction code TALYS.

1 Introduction

The concepts of the nuclear level density (NLD) and γ-ray strength function (GSF) are two essential tools for the statistical description of excited nuclei and their γ–decay, used in numerous large-scale astrophysical calculations of abundances of elements in the universe [1]. One of the widely used experimental techniques, the Oslo method [2] allows for the simultaneous extraction of both these nuclear properties from γ-decay of excited states within the quasi-continuum populated in light particle induced reactions. One of the key assumptions adopted in the Oslo method is the validity of the generalized Brink-Axel hypothesis (gBA) stating an independence of the GSF of excitation energies, spins and parities of initial and final excited states involved. The same assumption is usually adopted for calculations of different reaction rates, including those of astrophysical importance, using the nuclear reaction codes, *e.g.* TALYS [3].

Comparison of the Oslo method GSFs with strengths obtained with other experimental methods allows to address the question on the validity of the gBA hypothesis below the neutron separation energy for studied nuclei. With this assumption being valid, the experimental GSFs and NLDs can be used to constrain the radiative neutron capture rates and Maxwellian-averaged cross sections (MACS), essential for the astrophysical s- and r-processes. In the present work, the Oslo method was applied to the ^{120,124}Sn isotopes to extract the GSFs and NLDs to be further constrained by other experimental techniques and exploited as the nuclear input for astrophysical purposes.

2 Nuclear level densities and γ-ray strength functions

The ^{120,124}Sn were studied in the ^{120,124}Sn $(p,p'\gamma)$ reaction with the 16 MeV proton beam, performed at the Oslo Cyclotron Laboratory. The setup comprises of the particle silicon tele-

*e-mail: maria.markova@fys.uio.no

EPJ Web of Conferences 275, 02007 (2023) https://doi.org/10.1051/epjconf/202327502007
11th European Summer School on Experimental Nuclear Astrophysics

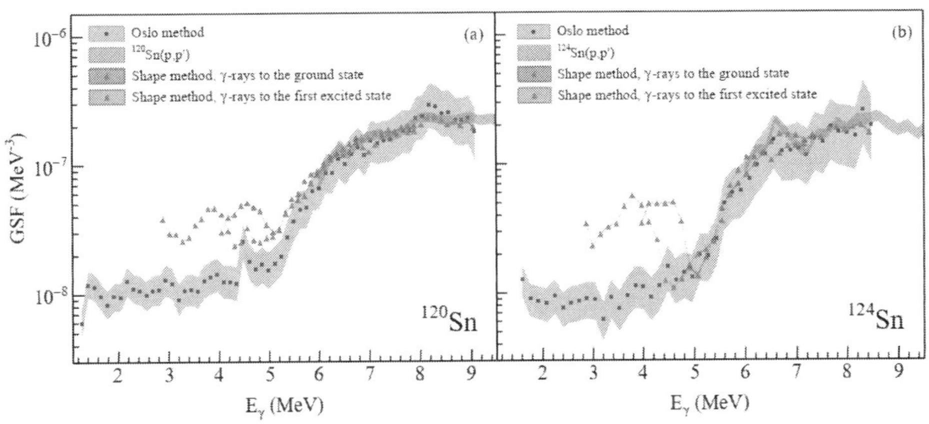

Figure 1. GSFs of ^{120}Sn and ^{124}Sn obtained with the Oslo method (blue band), the Shape method for γ-decays to the ground state (red band) and to the first excited state (green band) and the inelastic proton scattering experiment (orange band) [5].

scope SiRi placed in the backward position and the scintillator detector array OSCAR (see Ref. [2]). The setup is made to record the particle-γ coincidence data used as an input for the Oslo type of analysis. The raw coincidence data were processed to extract the primary matrix, consisting of the first generation photons in each cascade stemming from a nucleus at excited states below the neutron separation energy. The core idea behind the Oslo method is the decomposition of this primary matrix into the functional forms of the NLD and the GSF. However, these forms require external experimental data, such as low-lying discrete states, neutron resonance spacings and radiative widths, to constrain the slope and the absolute values of the NLD and GSF, *i.e.* to perform the normalization (see Ref. [2] for further details). Unfortunately, these normalization parameters are unavailable for ^{124}Sn and one has to rely on systematics of the total average radiative widths and NLDs at the neutron separation energies for other Sn isotopes. To account for errors of estimation of the normalization parameters from these systematics, the novel Shape method was additionally used to constrain the slope shared by the NLD and GSF [4]. This method allows to reconstruct the slopes of the strength from γ-decays to the ground and the first excited states in ^{120,124}Sn obtained in the same $(p, p'\gamma)$ reaction. Both the Shape and the Oslo method assume a predominant contribution of dipole transitions at sufficiently high excitation energies, which is supported by the experimental evidence [2], and, therefore, yield the dipole GSFs.

The comparison of the Oslo method with the Shape method results and the strengths deduced from spectra of relativistic Coulomb excitation in forward-angle inelastic proton scattering [5] is presented in figure 1. All GSFs agree quite well in shape within the estimated error bands above $\approx$ 5 MeV and below the neutron separation energy. The Shape method GSFs demonstrate an upward trend at low γ-ray energies due to the failure of some of the assumptions of the method at low energies. The details of the limitations of the Shape method are presented in Ref. [6]. This direct comparison of the γ-decay strength with the ground state photoabsorption strength suggests that the GSF is somewhat independent of the excitation energy below the neutron separation energy in accordance with the Brink-Axel hypothesis. An explicit study of the excitation energy dependence of the Oslo method GSF also reveals that the strengths for different initial and final excitation energies agree quite well within the error bands with the Oslo method result. The comparison of these results with the Shape method strengths above 5 MeV, where the Shape method can be considered applicable, also suggests some independence of the GSF of final spins $J = 0$ and $J = 2$ of studied transitions, corresponding to direct decays to the ground and first excited states.

The NLDs for ^{120,124}Sn are quite close in shape and absolute values to the previously studied NLDs for other even-mass Sn nuclei ^{116,118,122}Sn [6]. Moreover, the reduced Oslo method NLDs of 1^- states for ^{124}Sn agrees quite well within the error bars with the corresponding NLD from the Coulomb excitation experiment.

2.1 Maxwellian-averaged cross-sections

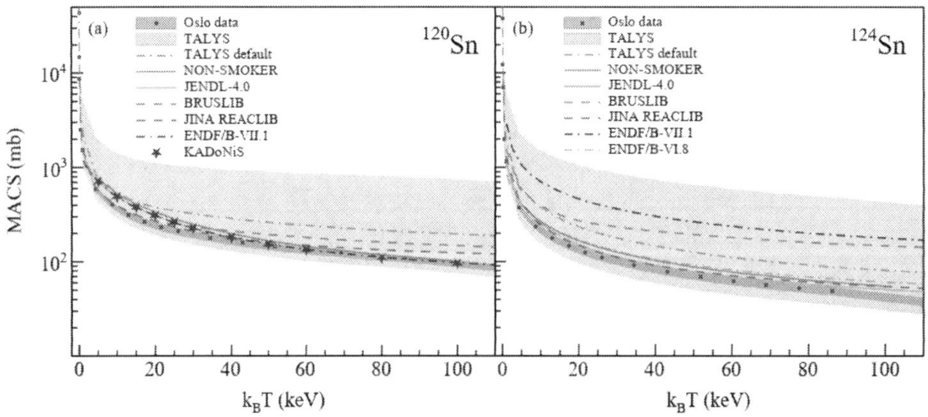

Figure 2. MACS of ^{120}Sn and ^{124}Sn obtained with the Oslo method input NLDs and GSFs compared to the span of TALYS predictions (light blue band), NON-SMOKER [7], JENDL-4.0 [8], BRUSLIB [9], JINA REACLIB [10], ENDF/B-VII.1 [11], ENDF/B-VI.8 [12], KADoNiS [13] libraries.

The experimental NLDs and GSFs were further used to constrain the MACS of the ^{119,123}Sn $(n,\gamma)^{120,124}$Sn reactions with the TALYS code. Both results are presented in figure 2 in comparison with the span of TALYS predictions, NON-SMOKER [7], JENDL-4.0 [8], BRUSLIB [9], JINA REACLIB [10], ENDF/B-VII.1 [11], ENDF/B-VI.8 [12] and KADoNiS [13] libraries. For both nuclei the experimentally constrained MACS deviate significantly from the default TALYS prediction and lie closer to the bottom par of the span of TALYS cross-sections. In is not surprising given large deviations of the TALYS NLD and GSF models from the experimental results that might reach up to several orders of magnitude for some excitation and γ-energies correspondingly. For ^{120}Sn, the MACS agree quite well with KADoNiS, ENDF/B-VII.1, JINA REACLIB and JENDL-4.0 values. The situation is different for ^{124}Sn , where the MACS from different libraries differ significantly from each other and from the experimental result, providing relatively low estimates of MACS values.

References

[1] S. Goriely, Phys. Lett. B **436**, 10 (1998)
[2] A. C. Larsen, *et al.*, Phys. Rev. C **83**, 034315 (2011).
[3] A. Koning, *et al.* , Nucl. Data Sheets **155**, 1 (2019)

[4] M. Wiedeking, *et al.*, Phys. Rev. C**104**, 014311 (2021).
[5] S. Bassauer, *et al.*, Phys. Rev. C **102**, 034327 (2020).
[6] M. Markova, *et al.*, Phys. Rev. C **106**, 034322 (2022).
[7] T. Rauscher and F. K. Thielemann, At. Data Nucl. Data Tables **75**, 1 (2000).
[8] K. Shibata, *et al.*, J. Nucl. Sci. Technol. **48(1)**, 1-30 (2011)
[9] Y. Xu, *et al.* , A&A **549**, A106 (2013).
[10] R. H. Cyburt, *et al.*, Astrophys. J., Suppl. Ser. **189**, 240 (2010).
[11] M. B. Chadwick, *et al.*, Nucl. Data Sheets **112**, 2887-2996 (2011).
[12] H. D. Lemmel, *et al.*, IAEA-NDS-100, Rev. **11** 1-32 (2001).
[13] Data extracted using the KADoNiS On-Line Data Service, http://www.kadonis.org, accessed 2022-08-05.

Laboratory magnetoplasmas as an ideal experimental environment for nuclear astrophysics β-decay studies

Eugenia Naselli[1,*]

[1]Istituto Nazionale di Fisica Nucleare - Laboratori Nazionali del Sud (INFN-LNS), Catania, Italy

Abstract. The PANDORA project proposes a new experimental approach aimed at using laboratory magnetoplasmas (which emulate some stellar conditions) as an environment for in-plasma β-decays investigations. In the superconducting PANDORA trap, a hot plasma containing a known concentration of β-decaying atoms can be confined and kept in dynamic equilibrium for weeks. The decay rate can be measured by detecting the γ-rays emitted by the daughter nuclei (through HPGe detector array) and correlated with the charge state distribution of radioactive ions and with the plasma thermodynamic properties using a multi-diagnostic system, whose tools and techniques are here presented.

1 Introduction

Magnetized plasmas in compact traps can become an experimental environment for studies of nuclear astrophysical interest. The PANDORA (Plasmas for Astrophysics, Nuclear Decay Observation and Radiation for Archaeometry) project [1] proposes a new experimental approach to measure in-plasma nuclear β-decays as a function of thermodynamic conditions of the environment. Theoretical models predict that the ionization state of the in-plasma isotopes can dramatically change the isotopes lifetimes [2], due to mechanism known as bound state β-decay [3], but only few experimental evidences have been collected up to now [3–5]. In the high-performance PANDORA plasma trap, a plasma is confined by multi-Tesla magnetic fields and resonantly heated by some kWs of microwave power in the 18-21 GHz frequency range. The plasmas can reach $n_e \sim 10^{11} - 10^{13} cm^{-3}, T_e \sim 0.1 - 100\, keV$ of electron density and temperature, respectively, and mimic some stellar-like environments, mainly in terms of the charge state distribution (CSD). Thus, radionuclides can be trapped in a dynamic equilibrium [6], maintaining an on-average locally stable density, temperature and CSD for weeks. An accuracy of about 10-15% (estimated using fast response diagnostics [7], i.e., soft-X ray spectroscopy [8]) can be achieved in terms of the maximal fluctuation of the main plasma parameters, which are deemed to be sufficient to get a good overall sensitivity in the measurement of the radioactive isotope decay. Details will be discussed in the next section. The overall structure of PANDORA setup (sketched in Fig. 1) consists of three main pillars: a) the *Magnetic Trap* [9]: a set of three superconducting coils and a hexapole for axial and radial plasma confinement; b) an *array of 14 HPGe detectors* [10] (overall photopeak efficiency 0.1-0.2%) to tag the in-plasma nuclear β-decays via the γ-rays emitted from the excited states of the daughter nuclei; c) the *plasma multi-diagnostics system* [7], consisting in a set of non-invasive diagnostic tools to locally characterize plasma thermodynamic properties.

*e-mail: eugenia.naselli@lns.infn.it

2 On-line investigation of ECR plasma thermodynamical properties

ECR plasmas are in n-LTE (non-local thermodynamic equilibrium) condition, thus the distribution function $f(\epsilon)$ is a convolution of three electron populations: the *hot* ($kT_e > 100$ keV), the *warm* (100 eV $< kT_e <$ 10 keV) and the *cold* (1eV $< kT_e <$ 100 eV) one.
The experimental aim is to conceive techniques to identify the influence of extrinsic factors coming from the external plasma-environment, which influence the decay constant λ of the trapped atoms, which should be in turn a function of the plasma electron density n_e and their energy distribution $f(\epsilon)$: $\lambda \equiv \lambda(n, f(\epsilon))$. Due to the dynamical equilibrium (n_e and $f(\epsilon)$ are kept constant within 10-15% over time) the number of expected decays N scales linearly with time as: $N(t)=\lambda(n_e, f(\epsilon))n_i V_p t$, being V_p and n_i the plasma volume and the isotope density.

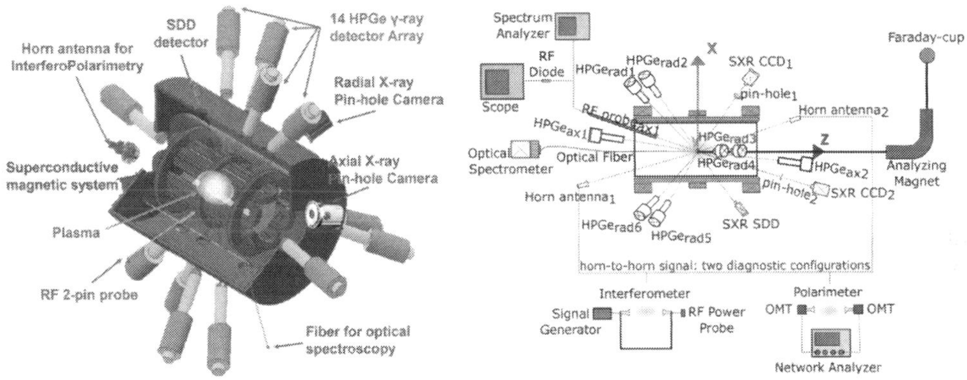

Figure 1. Rendering (left) and sketch (right) of the PANDORA setup with the systems used to measure in-plasma activity (14 HPGe detector array) and monitor plasma properties (multi-diagnostics system).

The core issue is to univocally determine n_e and $f(\epsilon)$, inherently fixing the CSD, to be correlated with the γ counting rate: this can be done identifying simultaneously plasma self emitted photons from microwaves to γ-rays. For a complete characterization of the overall $f(\epsilon)$, a multiplicity of tools and analysis methods becomes necessary [1, 7]. Due to the non homogenous nature of the plasma, this has to be done also using space and time-resolved techniques. A sketch of the multi-diagnostic system is shown in Figure 1-right and the main experimental techniques are summarized in the following. In particular, they are:

- **Volumetric X-ray Spectroscopy**: warm and hot electron populations can be studied through soft-X (Silicon Drift Detector SDD) and hard-X (HPGe detectors) volumetric spectroscopy. T_e and n_e can be estimated (with a typical relative error of around 10%-15%) by directly fitting experimental data, comparing the experimental spectrum with the theoretical one based on emissivity model [11–13]. The theoretical emissivity $J_{th}(h\nu)$ [11] is:

$$J_{th}(h\nu) = n_i n_e (Z\hbar)^2 \left(\frac{4\alpha}{\sqrt{6m_e}}\right)^3 \left(\frac{\pi}{kT_e}\right)^{\frac{1}{2}} e^{(-h\nu/kT_e)} \quad (1)$$

  thus, it is possible determining n_i and n_e (from spectrum's fit intersection with the vertical axis) and T_e (inversely proportional to the slope). Fig. 2.a) shows two hard X-ray spectra in which the measured temperature is 95 keV (blue) and 35 keV (red) respectively [8].

- **Spectrally-resolved soft X-ray Imaging**: the innovative pin-hole CCD camera technique [14], operating in single-photon-counted (SPhC) mode [15], allows spectrally-resolved imaging in the soft X-ray domain and, therefore, a deeper investigation of the parameters of the warm component. Figure 2.b) shows a SPhC soft X-ray energy-filtered image, corresponding to X-rays coming from plasma due to ionized Kα Argon lines (red) and X-rays coming from plasma chamber wall material, due to excited Ti (green) lines. Characteristic peaks associated to the emission from each material allow to investigate the spatial structure of the plasma [16] and confinement dynamics (plasma vs. losses X-ray emission) [17]. A model to link the experimental information to local plasma parameters is under development [12]. The profile of K_α emission intensity can be considered as: $I_{K_\alpha} \propto n_{Ar} n_e < \sigma_{K_\alpha} v >$, where σ_{K_α} is the K_α fluorescence cross section, the term $n_{Ar} n_e < \sigma_{K_\alpha} v >$ is the total reaction rate. An estimation of σ_{K_α} [12] and of the density and energy spatial distributions by simulations [13, 18] was given and the analysis for getting n_e and T_e is ongoing. In PANDORA two CCD pin-hole cameras will be used simultaneously (along the axial and radial lines) to estimate the plasma volume and thermodynamic conditions in each voxel, i.e. by a multi-pinhole tomography approach [19].

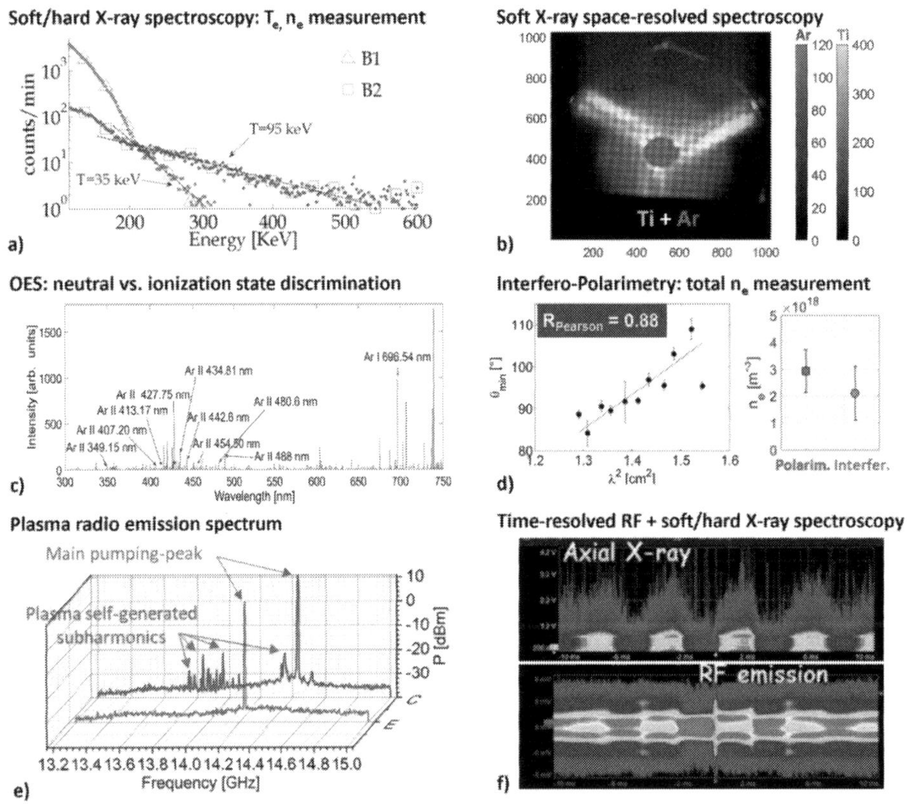

Figure 2. Overview of the main experimental results obtained through the multi-diagnostic system. a) Volumetric hard X-ray spectrum for two magnetic field profiles; b) Energy-filtered SPhC image of fluorescence X-rays coming from plasma (Ar) and from chamber walls materials (Ti); c) Optical emission spectrum of an Ar plasma; d) Faraday angle vs. probing wavelength in a polarimetric measurement (left) and total n_e measure by microwave interferometer and polarimeter (right); e) Plasma radio emission spectrum in a stable (green) and turbulent regime (red); f) time-resolved RF and hard X-ray spectra.

- **Optical Emission Spectroscopy (OES)**: the cold population can be characterized by OES measurements, comparing the experimental measured line ratio with the theoretical one estimated by means of a Collisional Radiative model [20–22], measuring the cold n_e and T_e with uncertainties that are less than 15% and 30%, respectively. Figure 2.c) shows a typical optical spectrum where it is possible to discriminate the neutral vs. the ionization state of an Ar plasma. For hydrogen plasma is possible to distinguish the molecular vs. atomic state and the lines of interest are the Balmer series (H_β/H_γ and H_α/H_β ratios). In perspective, the powerful spectrograph SARG (Spettrografo Alta Risoluzione Galileo), installed at LNS and reaching R=160.000 of resolution, will allow in-plasma on-line CSD measurement.

- **Microwave Interferometry and Polarimetry**: an interfero-polarimeter system, based on two high directive horn-antennas [23], allows to measure the total line-integrated density n_e. The polarimetric measurement is based on the evaluation of the Faraday rotation angle θ_{Far} of the polarization plane of a probing wave crossing the plasma, which is proportional to the square of the wavelength λ, to n_e (measured by fitting) and to magnetic field B (known), see fig. 2.d)-left. The equation is shown in the follow, where e, m, c, L are the charge and mass of electron, the speed of light and the plasma chamber length, respectively:

$$\theta_{Far} \propto \left(\frac{e^3}{2\pi m^2 c^4} \int_0^L n_e B dz \right) \lambda^2 \tag{2}$$

The interferometric n_e measurement is based on the phase-shift induced by the plasma refractive index. Results obtained by interferometry and polarimetry are in good agreement [24] (fig. 2.d-right), with errors respectively of 50% and 27%. Further improvements are expected by a new approach under development, based on a superheterodyne scheme to detect the Lissajous figure of the probing RF signals crossing the plasma [23], and in perspective by means of the microwave imaging profilometry to also measure the n_e profile.

- **RF diagnostics - plasma RF emission**: plasmas have to be maintained stable for weeks, avoiding onset of kinetic turbulences. Plasma stability can be monitored by means of a multi-pins RF probe connected to a Spectrum Analyzer, able to perform spectral analysis in the frequency domain [24]. Since plasma kinetic instabilities are characterized by RF and X-ray bursts (respectively with timescales of ns and ms), plasma radio-emission can be used as signature of turbulences. Two typical plasma radio emission spectra, respectively for a stable (green) and turbulent (red) regime, are shown in Fig. 2.e). In the latter case, the sub-harmonics are well visible and allow to quantitatively measure turbulences. For this purpose, a new parameter I_S was semi-empirically defined as follow [25]:

$$I_S = \left(\int_{13GHz}^{15GHz} \frac{dP(f)}{df} df - P_{mp} \right) (1 + w(N_{sub} - 1)) \tag{3}$$

where P_{mp} is the integral of the power of the main peak of pumping frequency, N_{sub} the number of sub-harmonics and w a weight factor. Since the instability strength is related both to the amplitude of the sub-harmonics and to their frequency spread, I_S was defined accordingly. It was calculated considering the amplitude (integral of the power) of RF plasma-self emitted signal, once subtracted the main pumping wave contribution, and then multiplied by a factor which considers the number of sub-harmonics N_{sub} with a proper weight factor w, which was optimized and set at 0.1. Experimental evidences demonstrated that I_S follows in a reasonable way what happens in the plasma in unstable conditions.

- **Time-resolved RF + soft/hard X-ray spectroscopy:** by connecting the RF probe with a diode and a scope it is possible to obtain high-resolution time-resolved (but totally integrated) power emitted from the plasma, using this value as trigger signal for X-ray detectors in order to perform volumetric X-ray spectroscopy [24]. Fig. 2.f) shows a comparison between the time-resolved RF spectra (bottom) vs. X-ray flux acquired by HPGe detector (top, reporting X-ray bursts in a turbulent regime of the plasma). This tool allows spectrally and timely characterization of turbulent plasma regimes, with a twice relevant outcome: i) a new "knob" for mastering plasma instabilities and maintain the plasma stable for weeks; ii) reproduce and study phenomena of astrophysics interest (kinetic turbulence occurring in astrophysical objects, i.e. Cyclotron Maser Instability [26]) in laboratory plasmas, measuring several properties even in the transient regimes. In perspectives, simultaneously space- and time-resolved spectroscopy will be performed by triggering the pin-hole system at the onset of plasma turbulences: this will allow to reach unprecedented capability of analysis of plasma dynamics, studying how the plasma shape and morphologies change during the time in a turbulence regime, also locally determining plasma parameters.

All diagnostics tools will operate simultaneously with a Faraday cup to measure the CSD.

3 Conclusion

PANDORA represents a promising experimental setup to verify, for the first time, the theoretical predictions on the dependence of lifetime on ion CSD. When the PANDORA setup will be ready (expected by 2024) the system will allow the simultaneous characterization of plasma properties, by a multidiagnostic system whose tools and techniques were described in this paper, and β-decay rates, also monitoring plasma stability and mitigating turbulences.

References

[1] D. Mascali et al., Universe **80**, 8 (2022)
[2] K. Takahashi et al., Phys Rev C **36**, 1522 (1987).
[3] F. Bosch et al., Phys. Rev. Lett. **77**, 5190 (1996).
[4] M. Jung et al., Phys. Rev. Lett. **69**, 2164 (1992)
[5] Y. Litvinov et al., Phys. Rev. Lett. **99**, 262501 (2007).
[6] R. Geller, ECR ion sources and ECR plasmas. Bristol, UK: J W Arrowsmith Ltd (1996).
[7] E. Naselli et al., J. Instrum. **14**, C10008 (2019)
[8] D. Mascali et al., Rev. Sci. Instrum. **87**, 095109 (2016)
[9] G. Mauro et al., Front. Phys. Sec. Nucl. Phys. **10**, 931953 (2022)
[10] E. Naselli et al., Front. Phys. Sec. Nucl. Phys. **10**, 935728 (2022)
[11] A. Gumberidze et al., Rev. Scient. Instr. **81**, 033303 (2010)
[12] B. Mishra et al., Phys. Plasmas. **28**, 102509 (2021)
[13] B. Mishra et al., Front. Phys. **10**, 932448 (2022).
[14] S. Biri et al., J. Instrum. **16**, P03003 (2021)
[15] E. Naselli et al., Condens. Matter **7**, 5 (2022)
[16] E. Naselli et al., J. Instrum. **17**, C01009 (2022)
[17] D. Mascali et al., Plasma Phys. Control. Fusion **64**, 035020 (2022)
[18] A. Galatà et al., Front. Phys. **10**, 947194 (2022)
[19] T. Sasaya et al., Sci. Rep. **7**, 5742 (2017)
[20] A. Pidatella et al., Front. Phys. Sec. Nucl. Phys. **10**, 931744 (2022)
[21] U. Fantz et al., Plasma Sources Sci. Technol. **15**, S137 (2006)
[22] U. Fantz et al., Nucl. Fusion **46**, S297 (2006)
[23] G. Torrisi et al., Front. Astron. Space Sci. **9**, 949920 (2022)
[24] D. Mascali et al., Rev. Scient. Instr. **93**, 033302 (2022)
[25] E. Naselli et al., Plasma Sources Sci. Technol. **28**, 085021 (2019)
[26] G.Q. Zhao et al., The Astrophysical Journal **822**, 58 (2016)

Direct measurement of the ^{26}Si(α, p)^{29}P reaction at CRIB for the nucleosynthesis in the X-ray bursts

Kodai Okawa[1,*], *Minju* Kim[2], *Kyungyuk* Chae[2], *Seiya* Hayakawa[1], *Satoshi* Adachi[3], *Soomi* Cha[4], *Thomas William* Chillery[1], *Tatsuya* Furuno[3], *Gyungmo* Gu[2], *Shutaro* Hanai[1], *Nobuaki* Imai[1], *David* Kahl[5], *Takahiro* Kawabata[3], *Chanhee* Kim[2], *Dahee* Kim[4], *Sohyun* Kim[2], *Shigeru* Kubono[6], *Minsik* Kwag[2], *Jiatai* Li[1], *Nanru* Ma[1], *Shin'ichiro* Michimasa[1], *Uyen* Nguyen Kim[2], *Duy* Nguyen Ngoc[2], *Kohsuke* Sakanashi[3], *Hideki* Shimizu[1], *Oana* Sirbu[5], *Hidetoshi* Yamaguchi[1], *Rin* Yokoyama[1], and *Qian* Zhang[1]

[1]Center for Nuclear Study, the University of Tokyo
[2]Department of Physics, Sungkyunkwan University
[3]Department of Physics, Osaka University
[4]Center for Exotic Nuclear Studies, Institute for Basic Science (IBS)
[5]Extreme Light Infrastructure Nuclear Physics (ELI-NP)
[6]RIKEN Nishina Center

Abstract. Several (α, p) reactions with radioactive-ions (RI) in the αp-process are important to characterize X-ray bursts. However, some of them do not have sufficient experimental data, and the ^{26}Si(α, p)^{29}P reaction is one of such reactions. We performed a direct measurement of the reaction in inverse kinematics with a thick target at the CNS RI beam separator (CRIB). We used a multiplexer circuit, Mesytec MUX, to acquire data from many channels of silicon detectors. In this experimental setup, a resonant elastic scattering was measured simultaneously. The details of the experimental conditions and the preliminary results of the analysis are discussed.

1 Introduction

The ^{26}Si(α, p)^{29}P reaction is one of the most influential reactions in X-ray bursts (XRBs). In the XRBs, the nucleosynthesis proceeds mainly by the rp-process, the rapid proton capture process. However, in the high temperature condition of XRBs (1–3 GK), the αp-process, the sequence of (α, p) and (p, γ) reactions, competes with proton capture reactions in the region of the nuclei that have atomic numbers of 20–50. The main observable of the XRBs is its light curve and Cyburt *et al.* [1] studied the sensitivity of the nuclear reactions on the light curve. This work classified the ^{26}Si(α, p)^{29}P reaction as "Category 1" in the single-zone model, which contains the reactions that have the greatest impact on the light curve. In addition, according to Parikh *et al.* [2], the reaction rate has some sensitivity on the galactic ^{26}Al abundance. The 1.808 MeV γ-ray line from ^{26}Al provides important information for the nucleosynthesis in the universe. Therefore, the nuclear data of the ^{26}Si(α, p)^{29}P reaction are essential for understanding the XRBs.

*e-mail: okawa@cns.s.u-tokyo.ac.jp

However, experimental data for this reaction are still insufficient and no direct measurement have been reported. Its reaction rate has been evaluated once before by Almaraz-Calderon *et al.* [3] based on the resonance information from the ^{28}Si(^{3}He, n) and ^{32}S(p, t) reactions. However, only in a limited excitation energy region up to 12.04 MeV and resonance parameters just above the α-threshold (9.343 MeV) were not definitely determined. Therefore, we performed direct measurements of the ^{26}Si(α, p)^{29}P reaction and resonant elastic scattering in excitation energies above 12 MeV simultaneously.

2 Experiment and Result

The experiment was performed at the RI Beam Factory operated by RIKEN Nishina Center and Center for Nuclear Study (CNS), the University of Tokyo. We used the CNS Radioisotope Beam Separator (CRIB) [4]. As shown in Figure 1, a ^{24}Mg^{8+} primary beam at 7.55 MeV/u and 3 eμA was irradiated at the ^{3}He primary target at 250 Torr, and a ^{26}Si RI beam was produced by the ^{3}He(^{24}Mg,^{26}Si)n reaction. The ^{26}Si beam purity was typically 29.0% after the two dipole magnets and the Wien filter, and the beam hits the ^{4}He gas secondary target at 250 Torr and room temperature (0.053 mg/cm^{3}). The typical intensity of the ^{26}Si beam was 32 kcps, and the beam energy was 4.5 MeV/u just after the Wien filter.

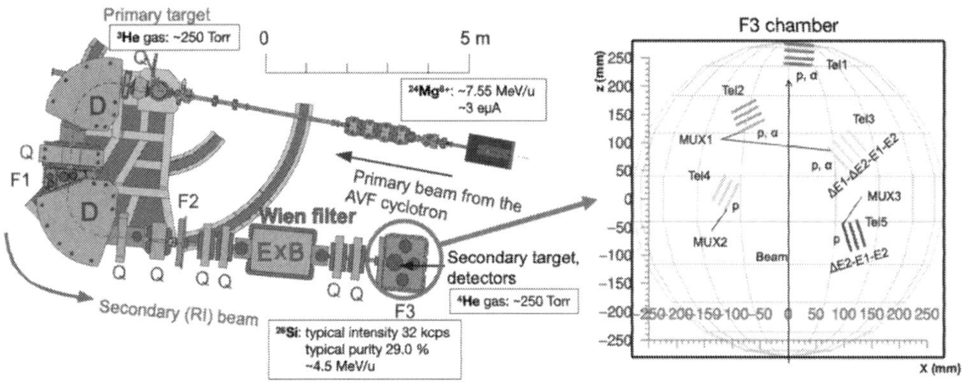

Figure 1. The left part shows the structure of CRIB facility and the parameters of the beam and the target. The right part shows the setup of the detectors in the target chamber. Note that the beam direction is from bottom to top. The three lower angle telescopes (Tel1, Tel2 and Tel3) consisted of two single-sided silicon strip detectors (SSSD) and two single-pad silicon detectors (SSD) and the two higher angle telescopes (Tel4 and Tel5) consisted of one double-sided silicon strip detector (DSSD) and two SSDs.

The thick target method [5] was used in this experiment in order to obtain the data of a wide range of scattering energy. The detectable range of the center-of-mass energy ($E_{\mathrm{c.m.}}$) was between 3.0 and 8.0 MeV. The longer the beam particles traveled to react, the smaller the $E_{\mathrm{c.m.}}$ of the reaction became. Therefore, it was possible to measure a wide energy range without changing the beam energy. The reaction products of p and α from the ^{26}Si(α, p)^{29}P reaction and the elastic scattering were identified by silicon ΔE-E telescopes. The configuration of the detectors is shown in Figure 1.

Compared to previous CRIB experiments, the number of the channels was larger in the present work and the original standard circuits could not handle all the channels. Therefore, multiplexer modules, Mesytec MUX [6], were introduced into the CRIB circuit for the first time. The inputs of the module are the 16 signals from a silicon strip detector and the outputs are two energies, two positions and one trigger (timing), and the module can handle up to two simultaneous hits. Three MUX modules with 32 input channels were used, and 192 channels were reduced to 30 channels in total including the energy and timing signals. The strip number can be known from the position output for each event, therefore it can process events in the same way as standard circuits. An energy resolution of the MUX circuits was equivalent to that of the original standard circuits.

Figure 2 shows the particle identification of one of the ΔE-E silicon telescopes (Tel2). This telescope consisted of 20 μm SSSD (dE1), 300 μm SSSD (dE2), 1494 μm SSD and 1485 μm SSD. The locus of the left and right panel in Figure 2 shows α and p respectively. These plots show that the ^{26}Si(α, p)^{29}P and elastic scattering were successfully detected in this telescope. A futher analysis is ongoing.

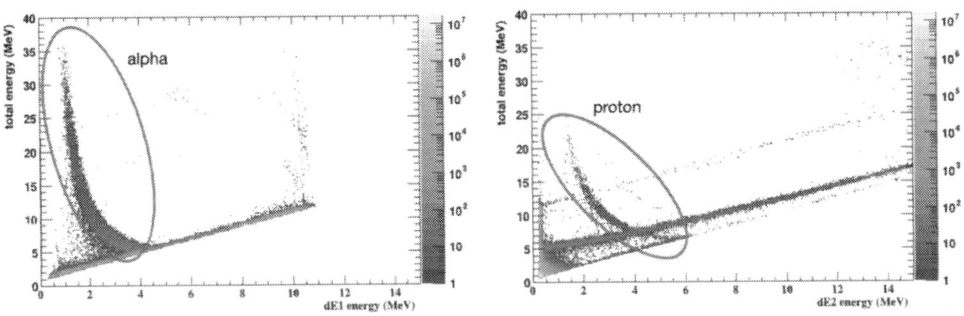

Figure 2. The particle identification using ΔE-E method. This data are from one of the lower angle telescope (Tel2). The left panel shows the total energy vs. the energy detected at dE1 and the right panel shows the total energy vs. the energy detected at dE2. The left and right loci indicate the α and p, respectively.

3 Summary

We performed direct measurements of the ^{26}Si(α, p)^{29}P and the ^{26}Si + α resonant elastic scattering simultaneously at CRIB. Because many detectors were required in this experiment, the multiplexer modules, mesytec MUX, were used first in CRIB. We succeeded in handling a large number of channels in limited feedthroughs with almost the same energy resolution as the original standard circuits. By using the ΔE-E method, the reaction products of α and p were identified. The reaction cross section of these reactions will be obtained in a future analysis.

References

[1] R.H. Cyburt *et al.*, Astrophys. J. **830**, 55 (2016)
[2] A. Parikh *et al.*, Astrophys. J., Suppl. Ser. **178**, 110 (2008)
[3] S. Almaraz-Calderon *et al.*, Phys. Rev. C **86**, 065805 (2012)
[4] Y. Yanagisawa *et al.*, Nucl. Instrum. Methods Phys. Res., Sect. A **539**, 74 (2005)
[5] K. Artemov *et al.*, Sov. J. Nucl. Phys. **52**, 408 (1990)
[6] Mesytec GmBH & Co., datasheet of MUX-16/-32/-64, https://www.mesytec.com/products/datasheets/MUX-16.pdf

The $^{12}C + ^{16}O$ fusion reaction in carbon burning: study at energies of astrophysical interest using the Trojan Horse Method

A. A. Oliva[1,2], A. Tumino[2,3], N. Soic[4], P. M. Prajapati[2,5], L. Acosta[6], R. Alba[2], F. Barba[7], S. Cherubini[1,2], G. D'Agata[1,2], D. Dell'Aquila[4], A. Di Pietro[2], J.P. Fernandez[8], P. Figuera[2], D. Galaviz Redondo[7], L. Guardo[1,2], M. Gulino[2,3], F. Hammache[9], D. Jelavic Malenica[4], A.I. Kiliç[10], M. La Cognata[2], M. La Commara[11,12], L. Lamia[1,2,13], D. Lattuada[2,3], C. Maiolino[2], G. Manicò[1,2], M. Mazzocco[14,15], M. Milin[4,16], Ma Nanru[17], A. Nurmukhanbetova[18], D. Nurkic[4,16], S. Palmerini[19], T. Parascandolo[11], D. Pierroutsakou[11], R.G. Pizzone[2], R. Popocovski[4], G.G. Rapisarda[1,2], S. Romano[1,2,13], D. Santonocito[2], M.L. Sergi[1,2], A. Shotter[20], R. Spartà[1,2], A. Spiridon[20], L. Trache[21], N. Vukman[4], and H. Yamaguchi[17]

[1]Dipartimento di Fisica e Astronomia "E. Majorana", Univ. di Catania, Italy
[2]INFN-LNS, Laboratori Nazionali del Sud, Catania, Italy
[3]Facoltà di Ingegneria e Architettura, Università degli Studi di Enna "Kore", Enna, Italy
[4]Rudjer Boskovic Institute, Zagreb, Croatia
[5]Manipal Centre for Natural Sciences, Manipal, India
[6]Instituto de Física, Universidad Nacional Autónoma de México, Mexico City, Mexico
[7]Departamento de Física, Faculdade de Ciências da Universidade de Lisboa, Lisboa, Portugal
[8]University of Seville, Seville, Spain
[9]IPN-Orsay, IN2P3-CNRS, Universite Paris XI, Orsay, France
[10]Nuclear Physics Institute of ASCR, Rez near Prague, Czech Republic
[11]Dipartimento di Fisica, Università degli Studi di Napoli "Federico II", Napoli, Italy
[12]INFN-Sezione di Napoli, Napoli, Italy
[13]Centro Siciliano Fisica Nucleare e Struttura della Materia, Catania, Italy
[14]Dipartimento di Fisica, Università degli Studi di Padova, Padova
[15]INFN-Laboratori Nazionali di Legnaro, Legnaro, Italy
[16]Physics Department, Faculty of Science, University of Zagreb, Zagreb, Croatia
[17]Center for Nuclear Study, The University of Tokyo, Tokyo, Japan
[18]Nazarbayev University, Astana, Kazakhstan
[19]Dipartimento di Fisica e Geologia, Univ. di Perugia, Italy.
[20]School of Physics - The University of Edinburgh, Edinburgh, Scotland
[21]IFIN-HH, Bucharest-Magurele, Romania

Abstract. The carbon-burning process in massive stars mainly occurs via the $^{12}C + ^{12}C$. However, at temperatures higher than $10^9 K$ and considering the increased abundance of ^{16}O produced during the later stages of the helium-burning, the $^{12}C+^{16}O$ fusion can also become relevant. Moreover, $^{12}C+^{16}O$ also plays a role in the scenario of explosive carbon burning. Thus, the astrophysical energy region of interest ranges from 3 to 7.2 MeV in the center-of-mass frame. However, the various measurements of the cross-section available in the literature stop around 4 MeV, making extrapolation necessary. To solve this uncertainty and corroborate direct measurement we applied the Trojan Horse Method to three-body processes $^{16}O(^{14}N, \alpha^{24}Mg)^2H$ and $^{16}O(^{14}N, p^{27}Al)^2H$ to study the $^{12}C(^{16}O, \alpha)^{24}Mg$ and $^{12}C(^{16}O, p)^{27}Al$ reactions in their entire energy region of astrophysical interest. In this contribution, after briefly describing the method used, the experiment and the preliminary phases of the data analysis will be presented and discussed.

1 Introduction

Carbon burning is one of the main advanced stages in the evolution of massive stars (M > 8·M$_\odot$) and type Ia supernovae [1, 2]. The main process of this stage is the ^{12}C +^{12} C fusion that ignites first during core and shell carbon burning. Indeed, among all the possible processes that can take place in the ashes of helium burning, this process has the lowest value for the Coulomb barrier. However, in the final phases of carbon burning, the abundance of ^{16}O nuclei is significantly higher [3, 4] than that of ^{12}C nuclei. Moreover, considering that also the temperature increases in these final phases, there is the possibility for other reactions with higher Coulomb barriers value to occur. It has been suggested[5] that the ^{12}C +^{16} O reaction temperatures higher than 10^9 K can have a significant impact on carbon burning. The latter is also of fundamental importance for Type Ia supernovae, since ^{12}C +^{12} C fusion is supposed to be its main energy source. Also in this case, where the relevant temperatures are around $3.6\cdot 10^9$ K, ^{12}C+^{16}O could also have a significant effect, as demonstrated in recent studies [6]. Thus, to define the importance of the ^{12}C+^{16}O fusion in all the aforementioned environments, it is necessary to precisely study this process in the energy range of astrophysical interest, namely between 3 and 7.2 MeV in the center of mass (cm).

In the literature there are several measurements of the ^{12}C +^{16} O fusion, however, they stop around 4 MeV, thus not covering all the Gamow region down to 3 MeV. This means that extrapolation must be used[7]: since this procedure strongly depends on the model used, on unforeseen nuclear effects and on the environment considered [8], results between different extrapolation at 3 MeV can vary up to two orders of magnitude [8]. Therefore, a new measurement also covering the energy region below 4 MeV could solve this uncertainty. In the aforementioned astrophysical scenarios, the ^{12}C+^{16} O fusion proceeds through the following reaction channels: ^{12}C(^{16}O, α)^{24}Mg (Q = 6.77 MeV), ^{12}C(^{16}O, p)^{27}Al (Q = 5.17 MeV) and partially ^{12}C(^{16}O, n)^{27}Si (Q = - 0.424 MeV). Reactions such as ^{12}C(^{16}O, 2α)^{20}Ne, similar to the case of ^{12}C +^{12} C fusion, are hindered by the presence of the Coulombian barrier in the exit channel [7, 11].

2 New THM Experiment

2.1 The Trojan Horse Method

The Trojan Horse Method (THM) is based on the well-established quasi-free break-up reaction theory (QF). To evaluate the cross-section of the two-body reaction of interest $x + A \to b + B$, the method studies an appropriate three-body reaction $a + A \to b + B + s$ where the nucleus a is assumed to possess a well-clustered state with a $x + s$ configuration. In this formalism, a is referred to as the Trojan Horse (TH) nucleus, the x cluster is called the participant, while s is called the spectator. Indeed, under the QF kinematic conditions, the momentum transfer to the cluster s is negligible acting as a spectator of the interaction between x and A [9]. Since the break-up of the TH nucleus occurs in the nuclear field between a and A, one of the advantages of the method is that the two-body reaction is not hindered by any barrier effects, both Coulombian and centrifugal, or effected by the electron screening [9]. Other peculiar advantages of the method are that with a fixed beam energy it is possible

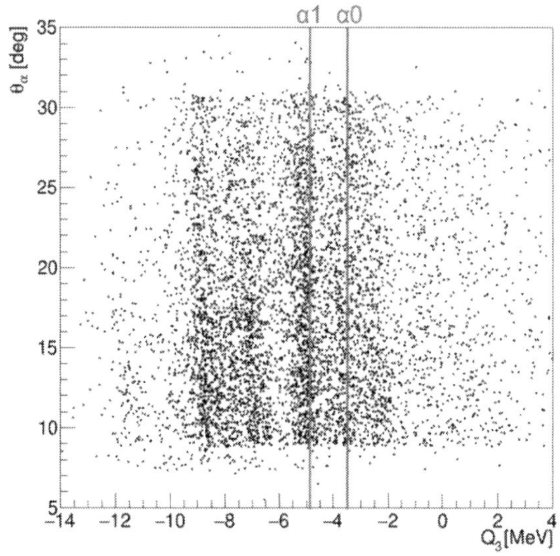

Figure 1. Detection angle of the alpha particle vs experimental Q-value for the $^{16}O(^{14}N, \alpha^{24}Mg)^2H$ reaction channel. Black dots represent the experimental data, while red lines represent the theoretical value for both the ground state ($\alpha 0$) and the first excited state ($\alpha 1$). From Ref. [12] with kind permission of Società Italiana di Fisica

to study a wide energy range in the center of mass frame between x and A, by taking advantage of the intercluster motion, and that it possible to study the sub-threshold energy region. A detailed explanation of the physics can be found in the already mentioned Ref. [9] and references therein.

2.2 Experimental Setup

The $^{12}C(^{16}O, \alpha)^{24}Mg$ and $^{12}C(^{16}O, p)^{27}Al$ reactions have been studied at astrophysical energies by applying the THM to the $^{16}O(^{14}N, \alpha^{24}Mg)^2H$ and $^{16}O(^{14}N, p^{27}Al)^2H$ three-body reactions. ^{14}N was chosen as the TH nucleus since it has a ^{12}C + d cluster structure [10] which was already successfully employed in a THM measurement [11]. Thus, a ^{14}N beam was accelerated by the Van de Graaff tandem of the Laboratori Nazionali del Sud (LNS-INFN) in Catania at an energy of 33.7 MeV to impinge on a 458 μg/cm^2 WO_3 target. Four telescopes, each of which was made up of three silicon detectors, were deployed to correctly detect and identify the particles in the exit channel. The first stage of each telescope consisted of a 35 μm Position Sensitive Detector (PSD), the second one of a 100 μm PSD, and the third one of a 1500 μm pad detector. The pad detector was placed to collect protons with high energy that can escape the second detector; meanwhile, the first stage was used in combination with the second stage to perform the $\Delta E - E$ technique for particle identification. The telescopes were placed to cover the QF angular range from 7° to 30° and from 45° to 68° which was obtained by means of a Monte Carlo simulation. Moreover, the setup was symmetrically doubled with respect to the beam axis in order to increase the statistics.

2.3 Data Analysis

The second-stage detectors were calibrated both in energy and position with devoted experimental runs using an eight-peak alpha source, elastic scattering of the ^{14}N on $^{1,2}H$ and the

^{2}H(^{14}N, α)^{12}C) reaction. Since the PSD does not have a segmentation, it was covered with a shield with equally spaced slits placed at known angles in order to perform the angular calibration. The third-stage detectors were calibrated only in energy. The first step of a THM analysis consists of precisely selecting the exit channel of the events: by applying the ΔE-E technique to the data collected from each telescope it is possible to perform the desired particle identification and subsequently select only the data corresponding to a coincidence event from a deuteron and either an alpha particle or a proton. The heavy nucleus in the exit channel was not detected, therefore the experimental Q-value was reconstructed by assuming the presence of an undetected ^{24}Mg or ^{27}Al. This assumption was then verified by comparing the experimental Q-value with the theoretical one: as it can be seen in Fig. 1, for the ^{16}O(^{14}N, α^{24}Mg)^{2}H reaction, it is indeed possible to assess the presence of the desired reaction process.

3 Conclusion and future perspectives

In this brief contribution, we show the preliminary results of the data analysis that demonstrate the correct execution of the experiment and the presence, among the data collected, of the ^{16}O(^{14}N, α^{24}Mg)^{2}H reaction channel of interest. The following steps of this analysis will involve the evaluation of the moment distribution of the spectator, to verify the presence of the QF mechanism, and the extraction of the cross section of the two-body reaction of interest. The same procedure will be then applied to the other reaction channel, the ^{16}O(^{14}N, p^{27}Al)^{2}H and, finally, the reaction rate of the ^{12}C +^{16} O process will be evaluated.

References

[1] Pignatari M. et al., Astrophys. J. **762**, 31 (2013)
[2] Chieffi A. et al., Astrophys. J. **916-2**, 79 (2021)
[3] Weaver T. A. et al., Phys. Rep.**227**, 65 (1993)
[4] Buchmann L., Astrophys. J. **468**, L127 (1996)
[5] Woosley S.E. et al., Phys. Rev. Lett. **27**, 213 (1971)
[6] Martínez-Rodríguez H. et al., Astrophys. J. **843**, 35 (2017)
[7] Fang X. et al., Phys. Rev. C **96-4**, 045804 (2017)
[8] Jiang C. L. et al., Phys. Rev. C **75**, 015803 (2007)
[9] Tumino A. et al., Annu. Rev. Nucl. Part. S. **71-1**, 345–76 (2021)
[10] Zurmühle R. W. et al., Phys. Rev. C **49**, 2549 (1994)
[11] Tumino A. et al., Nature **557-7707**, 687 (2018)
[12] Oliva A. A. et al., Il Nuovo Cimento C **45-5**, 119 (2022)

Sub-Coulomb barrier penetration for a ^{6}Li with a clustered and deformed ground-state

Salvatore Simone Perrotta[1,2,3,4,*], *Lorenzo* Fortunato[4,5], *José Antonio* Lay[2,6], and *Maria* Colonna[3,1]

[1]Università degli Studi di Catania – Catania, Italy
[2]Universidad de Sevilla – Seville, Spain
[3]Istituto Nazionale di Fisica Nucleare, Laboratori Nazionali del Sud – Catania, Italy
[4]Università degli Studi di Padova – Padua, Italy
[5]Istituto Nazionale di Fisica Nucleare, Sezione di Padova – Padua, Italy
[6]Instituto Interuniversitario Carlos I de Física Teórica y Computacional – Seville, Spain

Abstract. The penetrability of the Coulomb barrier of ^{6}Li by a proton is studied using a quantum cluster model. We focus on the role of quadrupole deformations in the nucleus ground-state, in terms of which a ^{6}Li−p form factor with tensor components is computed. We find that the diagonal part of the tensor term reduces the average barrier penetrability of the system. However, the tensor interaction due only to the mechanism studied at present is very small, regardless of the specific adopted construction, and yields negligible effects.

1 Introduction

The dynamics of Coulomb-barrier penetration plays a primary role in nuclear reactions taking place at collision energies of astrophysical interest [1, ch. 2]. The process can also be relevant with regards to the "electron screening problem" [2], the excess of enhancement observed in the cross-section of several nuclear reactions of astrophysical interest, measured in fixed-target experiments, with respect to the expected effect of the electrons in the projectile and target. Cluster models have been found to represent an excellent framework for the study of reactions involving light nuclei. In the context of barrier-penetration phenomena, the implications of such a model were explored in ref. [2] within a classical formulation.

In this work, we represent the structure of the ^{6}Li nucleus through a quantum two-cluster non-spherical wave-function. The projectile-target tensor interaction arising from such description, of which we aim to study the properties, appears to be a more microscopical analogous of the coupling interaction that is sometimes employed to study the impact of nuclear deformation on barrier penetrability in heavy-ion fusion reactions [3]. Additional details and previous work on this topic can be found in refs. [4, 5].

2 Description of the model and input ingredients

Let $\left|\mathrm{Li}_{1,M}\right\rangle$ be the ^{6}Li 1^+ ground-state with spin projection M. This is modelled as an inert-cluster-model bound state of α and deuteron, whose wave-function is written, as in ref. [5,

*e-mail: perrotta@lns.infn.it

eq. 1], as the superposition of s- and d-wave components. In refs. [4, 5] these components were constructed phenomenologically and combined with weights adjusted on data. Here, instead, the full wave-function was computed, using the FRESCO code [6], as the ground state of the α–d potential reported in ref. [7], including a central, a spin-orbit and a rank-2 tensor interaction. The volume term was fitted on a Woods-Saxon, whose depth was subsequently increased by 0.8 % to reproduce the α+d experimental binding energy. Such state corresponds to an electric quadrupole moment of −2.97 mb for the ^{6}Li, which does not agree satisfactorily with the experimental value of −0.81 mb [8] but still constitutes a relevant improvement with respect to the prediction of a central α–d potential (2.86 mb). Determining an inert-two-cluster-model state accurately reproducing all structure observables of interest appears to be challenging (for more discussion on this topic see refs. [4, 5] for a phenomenological construction and ref. [9] for a more microscopical one).

Given a set of phenomenological potentials for the interaction between the proton projectile and each ^{6}Li cluster, $V_{\alpha\mathrm{p}}$ and V_{dp}, if all couplings to ^{6}Li exited states are neglected, the ^{6}Li–p system can be described by just the ground-state form factor $V(M, M', \boldsymbol{R}) = \langle \mathrm{Li}_{1M'} | V_{\alpha\mathrm{p}} + V_{\mathrm{dp}} | \mathrm{Li}_{1M} \rangle$, where $\boldsymbol{R}$ is the distance between the centre of mass of projectile and target. Here, this was computed using the same potentials $V_{\alpha\mathrm{p}}$ and V_{dp} in ref. [5, sec. 3] and the present-work ^{6}Li wave-function. Even if the projectile-cluster potentials are central, the non-spherical ^{6}Li inter-cluster motion generates a tensor component in V, which can expressed as in ref. [4, sec. 5.2.2]. The qualitative features of the form factor computed here are similar to those found in ref. [5], but the weight of the tensor term is smaller (this is essentially because the norm of the ^{6}Li d-wave component is smaller here).

As in refs. [2, 4, 5], the barrier penetrability is evaluated approximately by neglecting the non-diagonal complex components of V (that is, selecting only $M = M'$) and computing the average, over each specific spin projection and orientation of the projectile-target displacement with respect to the quantization axis, of the WKB radial s-wave transmission coefficient (see e.g. ref. [3, eq. (3·60)] or [4, sec. 1.1, sec. 5]) for a central potential $V_{M,\theta}(R)$ equal to $V(M, M, \boldsymbol{R})$.

3 Results and conclusions

Figure 1 shows the average ^{6}Li–p penetrability computed as described in section 2, and compares it with the penetrability of the barrier corresponding to only the central part of $V(M, M, \boldsymbol{R})$, which is also equal to the average of V itself over all orientations and spin projections. The two calculations are practically indistinguishable. Having used a more microscopical description of the ^{6}Li cluster-model state, we thus confirm the result, mentioned in refs. [4, 5], that the diagonal part of the ^{6}Li–p tensor interaction generated solely by the non-sphericity of ^{6}Li ground state is too small to yield a measurable difference in the overall s-wave barrier penetrability. In order to strengthen the conclusion, it would be pertinent to add to the model the non-central components of the projectile-cluster potentials (in particular the tensor part of the deuteron interaction). Additionally, it may be important to improve the description of the barrier penetrability process, for instance taking into account the role of spin-flip processes allowed by the full form factor.

To investigate in greater detail the implications of this result, fig. 1 includes a numerical experiment (purple points) in which the tensor component of the form factor was rescaled by a factor of 20, without altering its shape (so that the average potential is unchanged): the associated penetrability is seen to be measurably smaller than in the previous cases. This is interesting, as the opposite result is frequently expected: nuclear deformations are often associated to fusion enhancement [3, sec. 3.1], and the average penetrability for a system

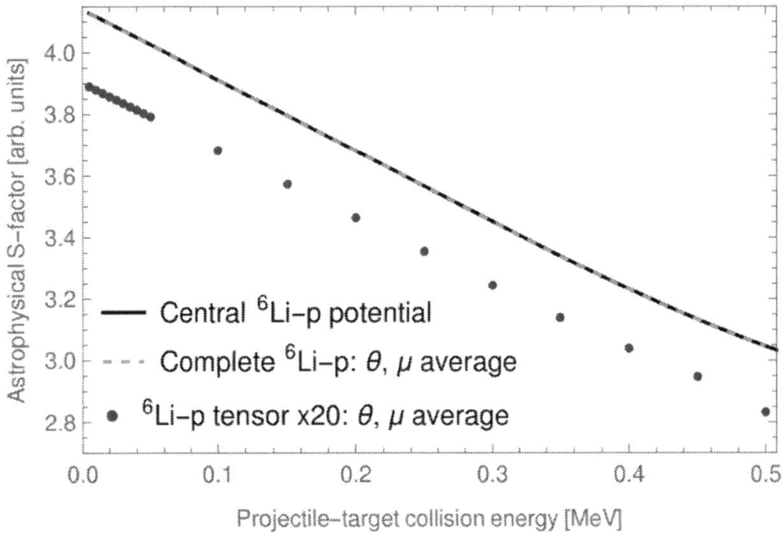

Figure 1. Astrophysical factor (see e.g. ref. [5, sec. 4] for the relevant definitions), in arbitrary units, for barrier penetration of the ^{6}Li−p system, computed as mentioned in section 2. The dashed orange line is the average over all configurations of the penetrability for the diagonal components of the present-work ^{6}Li−p form factor. The black solid line corresponds to only the central part of the same form factor. The purple points are the same as the dashed line but assuming a ^{6}Li−p tensor interaction 20 times bigger than computed in this work.

subject to atomic or plasma electron screening effects is always greater than the penetrability for the average screening potential [10, sec. 2.3]. We note that the outcome is sensitive to the shape of the perturbation to the potential barrier. Additionally, the present observation may in part be a consequence of the adopted approximations in the penetrability computation. As our preliminary results confirm, effects of nuclear origin can be non-trivial, and we plan to analyse them in greater detail in forthcoming works.

This project has received funding from the European Research Council (ERC) under the European Union's Horizon 2020 research and innovation program (grant agreement No. 714625).

References

[1] C. Iliadis, *Nuclear Physics of Stars*, 2nd edn. (Wiley-VCH, 2015)
[2] C. Spitaleri, C.A. Bertulani, L. Fortunato, A. Vitturi, Physics Letters B **755**, 275 (2016)
[3] K. Hagino, N. Takigawa, Progress of Theoretical Physics **128**, 1061 (2012)
[4] S.S. Perrotta, Ph.D. thesis, University of Catania and University of Seville (2022)
[5] S.S. Perrotta, L. Fortunato, J.A. Lay, M. Colonna, Il Nuovo Cimento C **45** (2022)
[6] I.J. Thompson, Computer Physics Reports **7**, 167 (1988)
[7] J.L. Gammel, B.J. Hill, R.M. Thaler, Phys. Rev. **119**, 267 (1960)
[8] *Nuclear electromagnetic moments*, `https://www-nds.iaea.org/nuclearmoments`
[9] A.C. Merchant, N. Rowley, Physics Letters B **150**, 35 (1985)
[10] S.S. Perrotta, A. Bonasera, Nuclear Physics A **989**, 168 (2019)

On the impact of compact binary merger ejecta opacity on Kilonova transient signals

Angelo Pidatella[1,*] for the PANDORA collaboration

[1]Istituto Nazionale di Fisica Nucleare - Laboratori Nazionali del Sud, Catania - Italy

Abstract. Returns of gravitational wave astronomy will largely benefit from the detection and identification of electromagnetic (EM) signatures to gravitational-wave sources. Kilonovae (KNe) are promising EM counterparts to compact binary mergers, offering to astronomers and nuclear astrophysicists a unique window to advance knowledge on the heavy-element nucleosynthesis and merger-driven mass ejection. However, extremely heterogeneous post-merging ejecta composition of both light- and heavy-r process nuclei, implies strong effects on the KNe light-curve identification due to the varying opacity of the system. Hence, large uncertainties on the r-process final abundance via spectroscopic analysis of KNe signals are still present, hardly fixed by theoretical models. Here we will present some peculiar features of KNe studies, focusing on the opacity issue, from the atomic and plasma physics perspectives. In this view, efforts have been made recently at INFN-LNS, trying to put constraints on plasma opacity of interest for early-stage KNe emission. We will present the experimental progress on the problem, including instruments and methods which could open an interdisciplinary approach to tackle astrophysical problems in laboratory plasmas.

1 Introduction

Matter ejected from the coalescence of compact binary objects, such as neutron-stars mergers, is thought to be among the major cosmic nuclei factories for the production of elements heavier than iron through the r-process nucleosynthesis [1–3]. Radioactive nuclei in the ejecta can feed thermal transients diffusing through the ejecta and the radiation can escape once when a translucent stage of the expanding ejecta is reached. These signals, known as *Kilonovae* (KNe), are electromagnetic (EM) counterparts of GW (Gravitational-Wave) events [4] and can last from days to weeks [5]. Their light-curves peak in different range of the EM spectrum [6], depending on many peculiar thermodynamic conditions of the ejecta, on the merging dynamics [5], and especially on the post-merging r-process nucleosynthesis pattern and final ejecta composition [7]. A good analysis of these signals becomes crucial for an efficient identification of GW signals, and for advancing knowledge on the equation of state of neutron stars. Moreover, KNe turn to be exceptional probes to give sounder constraints on the r-process nucleosynthesis yields. In the study of KNe, two fundamental inputs are requested [8]: (1) r-process yields (fixing the *heating rate*), (2) *opacity* (fixing the exchange of energy between plasma and radiation). The time-scale dynamics at which the KNe peak

*e-mail: pidatella@lns.infn.it

occurs depends on the opacity [2]. However, due to a strongly heterogeneous ejecta composition, opacity from theoretical models suffers from large uncertainty, with incomplete and uncertain atomic transition strength, and blending of plethora of transition lines, largely requesting for experimental data [7, 8]. The KN emission is reprocessed by atomic opacities (mainly bound-bound transitions) to optical and infra-red wavelengths. At the INFN-LNS, we propose an experimental design for plasma opacity measurements via spectroscopic techniques, in particular Optical Emission Spectroscopy (OES) analysis of emitting plasma with thermodynamic conditions resembling the astrophysical scenario, aiming at estimating the mean opacity for several light r-process metals abundant in the KNe diffusion stages [9, 10]. A feasibility study, to bound such experimental survey on typical laboratory plasmas used as plasma sources for accelerators, has been framed within the PANDORA project and research infrastructure, underlining that several physics cases are suitable for the study and envisaging the experimental design for the measurements [9]. The study opened the route to experimental measurements connecting nucleosynthesis (r-process), plasma physics, and nuclear astrophysics in the multi-messenger astronomy era. Experimental progresses have been done in terms of plasma parameters characterization via OES techniques as a function of several plasma source parameters [11], the summary of which is presented in the following section.

2 Experimental results

Before performing any of the planned opacity measurements, a robust plasma characterization is needed, attempting to reproduce and stabilize plasma conditions of interest. This must be done in respect of the several plasma trap parameters determining each experimental configuration. Since PANDORA is still under construction, we carried out these measurements on the Flexible Plasma Trap (FPT) [12] for only gaseous plasmas, as starting case, and adopted the methodologies employed in the OES to estimate the average electron energy (temperature k_bT_e), and density ρ_e along the line-of-sight of the detection system - see Fig. 1(a). A spectrometer coupled to a CCD detector has been used to acquire the H_2/Ar plasma emissions in the optical range (300-700 nm), tuning the gas pressure ($10^{-3} \div 10^{-2}$ mbar), microwave power ($40 \div 400$ W), frequency, and the confining magnetic field in the trap. By using the *line ratio method* [13] we were able to estimate k_BT_e and ρ_e for several stable configurations. Resulting data are summarily shown in Fig. 1(b), with more details on the experimental setup, configurations, and data analysis available in [11]. The peculiar distribution of the measured average electron temperatures (z-axis) and densities (y-axis) of plasmas explored as a function of microwave power (x-axis) is very promising in resembling the astrophysical conditions, desired to perform opacity measurements in the framework of the KNe. The strict requirement is on temperature, that must be at most few eV. Only few power/gas pressure configurations (H_2-Ar gas mixing, low RF power, $40 \div 200$ W, and pressure $\sim 10^{-2}$ mbar) allow these temperatures, providing a good track for the next future experimental setup. Plasma ejecta density condition of $10^{16 \div 18}$ m^{-3} are clearly fulfilled, independently on the RF power. The relative error calculated on the measurements is at most within the 50 % [11].

3 Conclusion

To conclude, we reported on the experimental plasma characterization whose work has been motivated by the possibility to reproduce in laboratory plasma ejecta conditions typical of a compact binary object post-merging phenomenon. The experimental survey served as a ground-state for the future experimental activity, with the aim to provide first-of-its-kind experimental opacity data useful for the KNe study. The measured data stressed the possibility to have controlled and monitored trapped plasmas with the desired thermodynamic and plasma parameters.

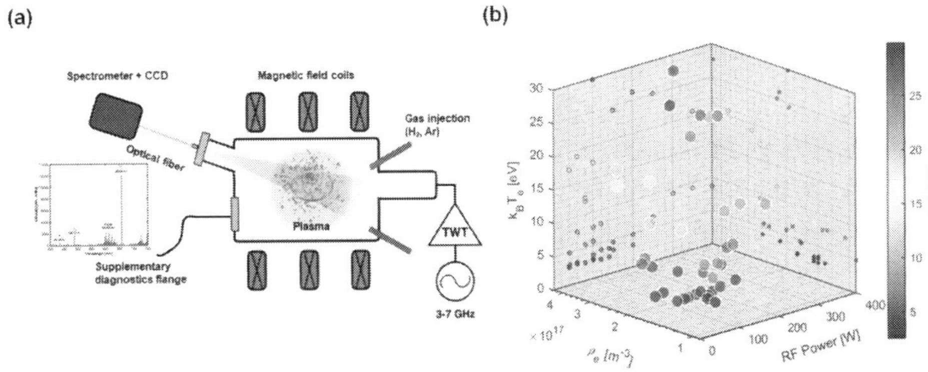

Figure 1. **(a)** Scheme of the experimental setup - further details in [11]. **(b)** Plasma electron energy (temperature, k_BT_e) [eV] (on z-axis) from OES measurements of H_2-Ar plasma emission in the FPT at the INFN-LNS. Data are displayed in the electron density ρ_e [m^{-3}] - microwave power P_{RF}[W] *xy*-plane. Projections onto *yz*-plane (k_BT_e vs. ρ_e), and *xz*-plane (k_BT_e vs. P_{RF}) are also shown. Data were collected for different experimental configurations in terms of power and gas pressure p_g. From projections, plasmas at low $k_BT_e \leq 3$ eV and $\rho_e \sim 1 \div 2 \cdot 10^{17}$ m^{-3} result only from H_2-Ar plasma, with $P_{RF} \leq 200$ W and high-pressure $p_g \sim 10^{-2}$ mbar.

Acknowledgments

The author thanks the INFN for the financial support through the project PANDORA Gr3 funded by 3rd Nat. Sci. Comm.

References

[1] L.X. Li, B. Paczyński, The Astrophysical Journal **507**, L59 (1998)
[2] B.D. Metzger, G. Martínez-Pinedo, S. Darbha, E. Quataert, A. Arcones, D. Kasen, R. Thomas, P. Nugent, I.V. Panov, N.T. Zinner, Monthly Notices of the Royal Astronomical Society **406**, 2650 (2010)
[3] O. Korobkin, S. Rosswog, A. Arcones, C. Winteler, Monthly Notices of the Royal Astronomical Society **426**, 1940 (2012)
[4] B.P. Abbott, R. Abbott, T.D. Abbott, F. Acernese, K. Ackley, C. Adams, T. Adams, P. Addesso, R.X. Adhikari, V.B. Adya et al. (LIGO Scientific Collaboration and Virgo Collaboration), Phys. Rev. Lett. **119**, 161101 (2017)
[5] D. Kasen, B. Metzger, J. Barnes, E. Quataert, E. Ramirez-Ruiz, Nature **551**, 80 (2017)
[6] I. Arcavi, G. Hosseinzadeh, D.A. Howell, C. McCully, D. Poznanski, D. Kasen, J. Barnes, M. Zaltzman, S. Vasylyev, D. Maoz et al., Nature **551**, 64 (2017)
[7] J. Barnes, D. Kasen, The Astrophysical Journal **775**, 18 (2013)
[8] M. Tanaka, D. Kato, G. Gaigalas, K. Kawaguchi, Mon. Not. Roy. Astron. Soc. **496**, 1369 (2020), 1906.08914
[9] A. Pidatella, S. Cristallo, A. Galatà, M. La Cognata, M. Mazzaglia, A. Perego, R. Spartà, A. Tumino, D. Vescovi, D. Mascali, Il Nuovo Cimento C **44**, 4 (2021)

[10] D. Mascali, D. Santonocito, S. Amaducci, L. Andò, V. Antonuccio, S. Biri, A. Bonanno, V.P. Bonanno, S. Briefi, M. Busso et al., Universe **8** (2022)
[11] A. Pidatella, D. Mascali, M. Bezmalinovich, G. Emma, M. Mazzaglia, B. Mishra, G. Finocchiaro, A. Galatà, S. Marletta, G. Mauro et al., Frontiers in Astronomy and Space Sciences **9**, 225 (2022)
[12] S. Gammino, L. Celona, D. Mascali, G. Castro, G. Torrisi, L. Neri, M. Mazzaglia, G. Sorbello, O. Leonardi, L. Allegra et al., Journal of Instrumentation **12**, P07027 (2017)
[13] U. Fantz, Plasma Sources Science and Technology **15**, S137 (2006)

Indirect measurement of the $(n,\gamma)^{127}$Sb cross section

Francesco Pogliano[1,*], *Ann-Cecilie* Larsen[1,**], *Frank Leonel* Bello Garrote[1], *Marianne* Møller Bjørøen[1], *Thomas* Kvalheim Eriksen[1], *Dorthea* Gjestvang[1], *Andreas* Görgen[1], *Magne* Guttormsen[1], *Kevin Ching Wei* Li[1], *Maria* Markova[1], *Eric Francis* Matthews[2], *Wanja* Paulsen[1], *Line* Gaard Pedersen[1], *Sunniva* Siem[1], *Tellef* Storebakken[1], *Tamás* Gabor Tornyi[1], and *Julian* Ersland Vevik[1]

[1]Department of Physics, University of Oslo, N-0316 Oslo, Norway
[2]Department of Nuclear Engineering, University of California, Berkeley, California 94720 U.S.A.

Abstract. Sensitivity studies of the *i* process have identified the region around ^{135}I as a bottleneck for the neutron capture flow. Nuclear properties such as the Maxwellian-averaged cross section (MACS) are key to constrain the uncertainties in the final abundance patterns. From the ^{124}Sn$(\alpha,p\gamma)^{127}$Sb reaction we are able to indirectly measure the nuclear level density and γ-ray strength function for ^{127}Sb using the Oslo method. From these two quantities we can calculate the MACS for the ^{126}Sb$(n,\gamma)^{127}$Sb reaction using the Hauser-Feshbach formalism, constrain its uncertainties and compare it to libraries such as JINA REACLIB, TENDL and BRUSLIB.

1 Introduction

The *s* and the *r* processes (standing for the slow and rapid neutron-capture processes, respectively) are the main mechanisms behind the creation of the elements heavier than iron in our universe [1]. The two nucleosynthesis processes produce different elemental abundance patterns, and these can be recognized in the observed abundances in stars [2]. This is the case for carbon-enhanced, metal-poor stars (CEMPs), which may be enriched in *r* process elements [3], *s* process elements, or both [4]. This last case is challenging to explain, as the two nucleosynthesis processes are thought to happen in different astrophysical sites, and CEMPs are thought to have formed before the interstellar medium from these sites could mix. A possible solution to this problem is the introduction of the *i* process. The *i* process stands for intermediate neutron-capture process was first proposed in 1977 by Cowan and Rose [5] and has neutron densities between those of the *s* and the *r* process. Simulations of this process can reproduce the abundances of CEMPs-s/r (CEMPs presenting elements characteristics to both the *s* and *r* process), and they identify the region around ^{135}I as a bottleneck [6]. However, one of the main sources of uncertainty lies in the correct estimation of nuclear properties such as the neutron-capture rates of the nuclei involved. By carrying out experimental studies in the ^{135}I region we are able to constrain these uncertainties. This region of the nuclear chart includes many unstable, neutron-rich nuclei, for whose it is very difficult to carry direct measurements of the (n,γ) reaction cross section. For this reason an indirect approach is

*e-mail: francesco.pogliano@fys.uio.no
**e-mail: a.c.larsen@fys.uio.no

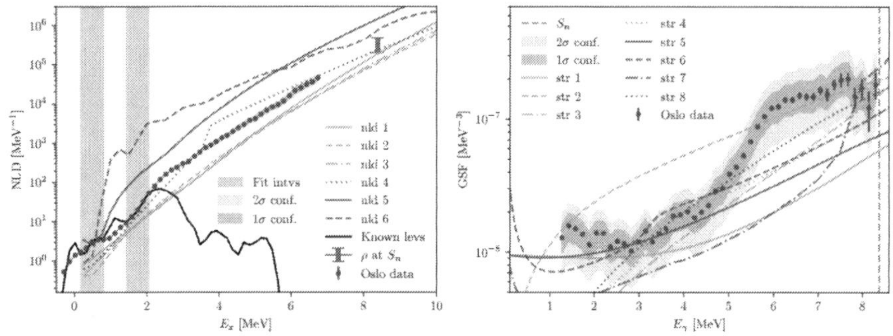

Figure 1. The NLD (left) and the GSF (right) for ^{127}Sb. `nld` and `str` indicate the model number obtained from TALYS 1.95 [15, 16]. The red bars in the left plot indicate the low excitation energy regions the NLD was normalized to (see Ref. [13] for details.)

used. With the Oslo method we are able to extract the nuclear level density (NLD) and the γ-ray strength function (GSF). These are statistical properties of the nucleus, the NLD being the "continuous" equivalent of counting energy levels, and the GSF of the reduced transition probabilities. These two quantities are the ingredients needed in the Hauser-Feshbach formalism (together with the optic model potential) to calculate the neutron-capture rate (see [7] and references therein.) Here we present the results from the ^{124}Sn$(\alpha, p\gamma)^{127}$Sb experiment. By analyzing the data with the Oslo method we obtain the NLD and the GSF of ^{127}Sb, part of the ^{135}I region, from whose we can calculate the $(n, \gamma)^{127}$Sb reaction rate.

2 The extraction of the NLD and GSF

The ^{124}Sn$(\alpha, p\gamma)^{127}$Sb reaction experiment was carried out at the Oslo Cyclotron Laboratory for a period of five days. A 24 MeV α-beam was impinged on a self-supporting ^{124}Sn target. In order to use the Oslo method, we were interested in particle-γ coincidences. For this reason the reaction data was collected using a silicon particle detector (SiRi [8],) and a $LaBr_3$ γ-ray detector (OSCAR, [9]). The particle detector consists of a thin front detector (ΔE) and a thicker back detector (E), and this allowed us to separate the different reaction channels and only select the (α, p) data. From the energy of the beam and that of the ejected proton, we were able to calculate the excitation energy the resulting ^{127}Sb is left in, and associate it to the coinciding γ-rays from its de-excitation to the ground state. These different γ-ray spectra for each excitation energy are plotted together in a matrix, called raw coincidence matrix. The Oslo method is a procedure for which we can obtain the NLD and the GSF of a nucleus starting from a raw coincidence matrix [10–12]. First the matrix has to be unfolded to correct for the detector response. Secondly, The first-generation matrix is obtained with a subtraction technique, where we only select the first γ-rays from each de-excitation. The NLD and GSF are used to describe the nucleus in the quasicontinuum excitation energy region, where the level density becomes so large ($\gtrsim$ 50 per MeV) that it is more useful to describe nuclear properties statistically. These two quantities were extracted from the first-generation matrix from this region using a χ^2 minimization algorithm explained in Ref. [12], and then normalized (for a detailed description of this procedure we refer to the main article about this experiment in Ref. [13]). The extracted NLD and GSF are shown in Figure 1. Here we see how the NLD follows an exponential curve compatible with the constant-temperature model. The GSF shows features such as the upbend at low energies, the pygmy resonance and

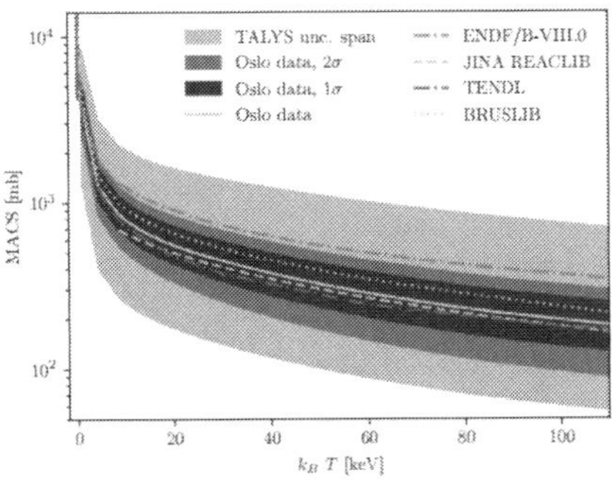

Figure 2. The constrained MACS uncertainty for the $(n,\gamma)^{127}$Sb reaction, compared to the uncertainty range from combining all the NLD and GSF theoretical models from TALYS.

possibly a small structure at ≈ 3 MeV. For each graph, all the available TALYS 1.95 [15, 16] models were plotted as comparison, showing a general disagreement to the experimental data.

3 The MACS

The Maxwellian-averaged cross section (MACS) is a quantity easily translatable to the neutron-capture rate [14] and is an important nuclear input parameter for nucleosynthesis simulations. Theoretical MACSs can be calculated with TALYS in the Hauser-Feshbach framework by choosing a NLD, a GSF and a optical model potential model. While the latter does not influence the MACS too much, combining all the models gives an uncertainty span of more than one order of magnitude (see Figure 2). The experiment data constrains this uncertainty and is compared to common libraries such as JINA REACLIB [17], BRUSLIB [18], TENDL [16] and ENDF/B-VIII.0 [19]. While the first three give predictions within 1σ uncertainty from the experimental result, the latter predicts values outside the 2σ uncertainty region.

References

[1] M. Burbidge et al., Rev. Mod. Phys. **29**, 547 (1957)
[2] M. Arnould et al., Physics Reports **450**, Pages 97-213 (2007)
[3] T. Hansen et al., ApJ **807**, 173 (2015)
[4] C. Sneden et al., Ann. Rev. of A&A **46:1**, 241-288 (2008)
[5] J. J. Cowan et al., ApJ **212**, 149-158 (1977)
[6] M. Hampel et al., ApJ **831**, 171 (2016)
[7] A.C. Larsen et al., PPNP **107**, Pages 69-108 (2019)
[8] M. Guttormsen et al., Nucl. Instrum. Methods Phys. Res. Sect. A **648**, 168 (2011)
[9] F. Zeiser et al., Nucl. Instrum. Methods Phys. Res. Sect. A **985**, 164678 (2020)
[10] M. Guttormsen et al., Nucl. Instrum. Methods Phys. Res. Sect. A **374**, 371 (1996)
[11] M. Guttormsen et al., Nucl. Instrum. Methods Phys. Res. Sect. A **255**, 518 (1987)
[12] A. Schiller et al., Nucl. Instrum. Methods Phys. Res. Sect. A **447**, 498 (2000)
[13] F. Pogliano et al. Phys. Rev. C **106**, 015804 (2022)
[14] C. Iliadis, N*uclear Physics of Stars* (Wiley-VCH, Weinheim, 2015) 147-150
[15] A. Koning et al., User manual, Tech. Rep., (2017)
[16] A. Koning et al., Nucl. Data Sheets **155**, 1 (2019)
[17] R. H. Cyburt et al., ApJ Suppl. Series **189**, 240 (2010)
[18] Y. Xu et al., Astron. Astrophys. **549**, A106 (2013)
[19] D. Brown et al., Nucl. Data Sheets **148**, 1 (2018)

^{13}C(α,n)^{16}O : The Source of Neutrons for the s-process main component

P. M. Prajapati[1,3*], *Mahin* Qureshi[2], *A.* Hingu[2], *R. G.* Pizzone[3], *M.* La Cognata[3], *S. V.* Suryanarayna[1,4], *Sachin* Shet[1], *S.* Mukherjee[5]

[1]Manipal Centre for Natural Sciences, Manipal Academy of Higher Education, Manipal - 576104, India
[2]Physics Department, Faculty of Science, M. S. University of Baroda, Vadodara - 390002, India
[3]Istituto Nazionale di Fisica Nucleare, Laboratori Nazionali del Sud, Catania - 95123, Italy
[4]Nuclear Physics Division, Bhabha Atomic Research Centre, Mumbai - 400085, India
[5]Manipal Institute of Technology, Manipal Academy of Higher Education, Manipal - 576104, India

Abstract. The ^{13}C(α, n)^{16}O reaction operates in the He intershell of low-mass (M < 4 $M_\odot$) AGB stars and it is the neutron source that allows the creation of the main component of the s-process elements. The reaction rate is then required to be well known in the energy range of astrophysical interest. Therefore, the ^{13}C(α, n)^{16}O reaction rate has been calculated in stellar like conditions using a nuclear-model based computer code TALYS. The results have been compared with available literature data and found to be in good agreement with experimental data and, evaluated data NACRE II as well.

1 Introduction

Asymptotic Giant Branch (AGB) stars play an important role in determining the galactic chemical evolution [1]. In their interiors, the main components of the s-process [2] is synthesized [3], namely the slow neutron capture process allows the production of several isotopes of elements from Sr to Pb. The s-process path runs along the stability valley due the competition between n-captures and beta decays [4]. The ^{13}C(α, n)^{16}O reaction is the main source of neutrons in low-mass AGB stars, which is active in the He intershell below the H shell [5]. It is well known that the ^{13}C pocket forms at the beginning of each interpulse [3] period in the He rich layers thanks to a special injection of a protons. When H shell re-ignites ^{13}C forms through the ^{12}C(p, γ)^{13}N (β^+)^{13}C and then starts capturing α particles and, as a consequence, releases neutrons [3]. This process provides relatively low neutron densities [$\approx 10^7 neutrons/cm^3$] for $10^4 - 10^5$ years each time. Starting from seed nuclei in the iron region, this neutron flux slowly builds up heavy elements along the line of stability. The^{13}C(α, n)^{16}O reaction takes place at 90 MK corresponding to effective energy range (i.e, Gamow window) of 140 - 230 keV. In few cases,^{13}C pocket also delivers neutrons at lower temperature down to 50 MK [1]. Fig.1 shows the normalized probability using Maxwell-Boltzmann distribution versus center-of-mass energy E for the^{13}C(α, n)^{16}O reaction at three different temperatures. It can be seen from Fig. 1 that slight variation in temperature shifts the Gamow window and peaks as well. In order to constrain this important nucleosynthesis process, the reaction rate of the ^{13}C(α, n)^{16}O neutron source needs to be known in the astrophysical energy

*e-mail: pareshkumar.p@manipal.edu

window. A detailed literature [6–12] survey indicates that the $^{13}C(\alpha, n)^{16}O$ reaction has been studied extensively. Most of the authors have reported cross-sections or astrophysical factors in stellar environments for this reaction. Only few of them have reported thermonuclear reaction rates. In this context, $^{13}C(\alpha, n)^{16}O$ reaction rate has been calculated using nuclear model based computer code TALYS [13] at the temperatures of astrophysical interest. The calculated reaction rates have been compared with some of the available experimental and evaluated data set NACRE II [14].

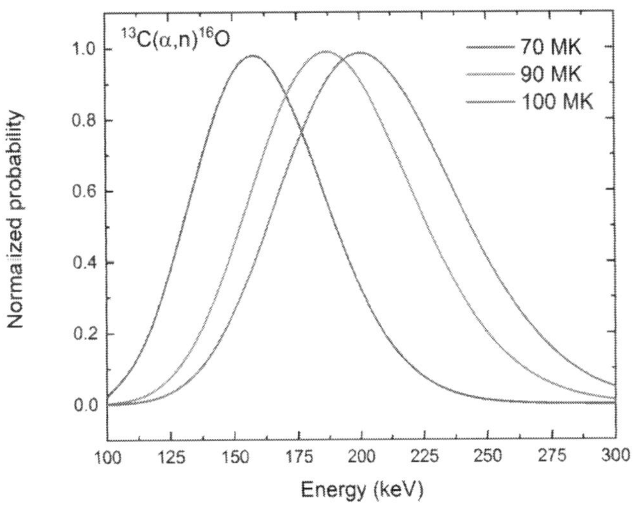

Figure 1. Maxwell-Boltzmann distribution versus energy E for the $^{13}C(\alpha, n)^{16}O$ reaction at three different temperature T.

2 Nuclear Model based Calculations

TALYS performs simulation of nuclear reactions that involve neutrons, protons, deuterons, tritons,^{3}He- and alpha- particles, in the 1 keV-200 MeV energy range and for target nuclides of mass 12 and heavier. It adopts Hauser-Feshbach (HF) statistical model [15] to estimate the reaction rates that are of astrophysical relevance. HF model relies on the fundamental assumption (Bohr hypothesis) that capture process occurs by means of the intermediary production of a compound nucleus (CN) that can reach a state of thermodynamic equilibrium. TALYS estimates the corresponding cross-section by the compound nucleus formula for the binary cross-section [16] as given in eq. 1,

$$\sigma^{\mu}_{\alpha\alpha'} = D^{comp}\pi\bar{\lambda}^2 \sum_{J=mod(I^\mu+s,1)}^{l_{max}+I^\mu+s} \sum_{\Pi=-1}^{1} \frac{2J+1}{(2I^\mu+1)(2s+1)} \sum_{j=|J-I^\mu|}^{J+I^\mu}\sum_{l=|j-s|}^{j+s}\sum_{j'=|J-I'|}^{J+I'} \sum_{l'=|j'-s'|}^{j'+s'} \delta_\pi(\alpha)\delta_\pi(\alpha') \frac{\langle T^J_{\alpha lj}(E_a)\rangle\langle T^J_{\alpha' l' j'}(E_{a'})\rangle}{\sum_{\alpha'',l'',j''}\delta_\pi(\alpha'')\langle T^J_{\alpha''l''j''}(E_{a''})\rangle} W^J_{\alpha lj\alpha' l' j'} \quad (1)$$

In the above equations, E_a, s, π, l and j represent the projectile energy, spin , parity, orbital and total angular momentum, respectively. The same symbols but labeled by a prime correspond to ejectile. The symbols $I^{\mu}, \Pi^{\mu}(I^{'}, \Pi^{'})$, represent the spin and parity of target nucleus, while J, and Π correspond to the spin and parity of compound system. A thermodynamic equilibrium holds locally to good approximation inside stellar interiors. Consequently, the energies of both target and projectiles, as well as their relative energies, obey a Maxwell-Boltzmann distribution corresponding to the temperature T. In such conditions, the astrophysical rate is obtained by integrating the cross-section given by eq. 1 over a Maxwell-Boltzmann distribution of energies at the given temperature T. The stellar rate per pair of particles in the entrance channel at temperature T, taking account of the contributions of various target excited state, is expressed in classical notation as given eq. 2.

$$N_A\langle\sigma v\rangle_{\alpha\alpha'}(T) = \left(\frac{8}{\pi m}\right)^{1/2} \frac{N_A}{(kT)^{3/2}G_I(T)} \int_0^{\infty} \sum_{\mu} \frac{(2I^{\mu}+1)}{(2I^{0}+1)} \sigma^{\mu}_{\alpha\alpha'} E(E) exp\left(-\frac{E+E_x^{\mu}}{kT}\right) dE \quad (2)$$

where k is the Boltzman constant, m is the reduced mass and N_A the Avogadro Number. Using the above adopted formalism in TALYS code, the rate of $^{13}C(\alpha, n)^{16}O$ reaction has been calculated. In statistical models, nuclear level densities are used at excitation energies where discrete level information is not available or incomplete. TALYS uses several models for level density, which range from phenomenological analytical expressions to tabulated level densities derived from microscopic models. In the present calculations, the level density corresponding to Constant Temperature + Fermi gas model introduced by Gilbert and Cameron [17] was used. In this model, the excitation energy range is divided into a low energy and high energy parts. Low energy part from 0 MeV up to matching energy E_M, where the so-called constant temperature law applies and high energy part above E_M, where the Fermi Gas model applies. The optical model parameters (OMP) are calculated using local and global parameterisations of Koning and Delaroche [18]. In addition, the alpha spherical local optical potentials of Avrigeanu et al [19] has been used. The theoretically calculated $^{13}C(\alpha, n)^{16}O$ reaction rate has been plotted along with a selected experimental data [11, 12] as well as the nuclear reaction rates compilation NACRE II as shown in Fig. 2. NACRE compiles a number of cross-section measurements and subsequently calculates a comprehensive reaction rates. Fig. 2 shows that the theoretically calculated reaction rates agrees very well with direct experimental data reported by G. F. Ciani et al [11] and NACRE II within the experimental errors. Moreover, Fig. 2 also report the data by O.Trippella et al, [12] whose have investigate the $^{13}C(\alpha, n)^{16}O$ reaction by combining two indirect approaches: the Asymptotic Normalization Coefficients [20] and Trojan Horse Method [21, 22]. In this case theoretical estimate matches the experimental data only at low and high temperature, and not in the central part of the investigated range. Therefore, we can conclude that TALYS provides a reasonable evaluation for the $^{13}C(\alpha, n)^{16}O$ reaction rates, even if it matches only experiential data directly measured. Keeping this warning in mind, we can state that TALYS might be used to estimate thermonuclear reaction rates in case of reactions where no experimental data are available (e.g. processes involving short lived nuclei interacting with neutrons) at the energies of typical astrophysical processes.

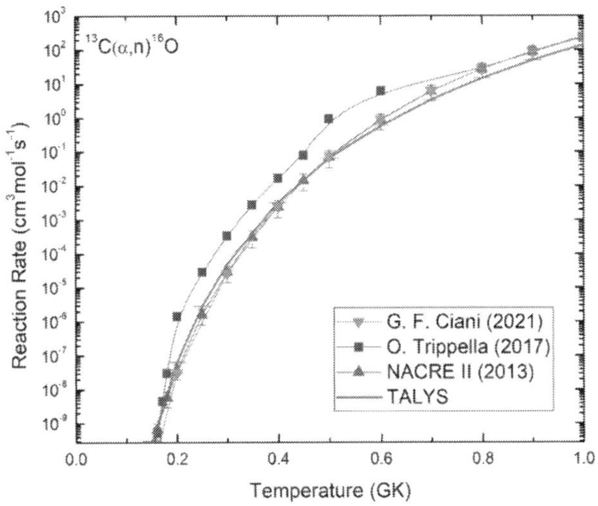

Figure 2. Theoretically calculated thermonuclear reaction rate for $^{13}C(\alpha, n)^{16}O$ reaction along with experimental data and NACRE II.

References

[1] M. Busso et al., Annu. Rev. Astron. Astrophys. **37**, 239 (1999)
[2] F. Herwig., Annu. Rev. Astron. Astrophys. **43**, 435 (2005)
[3] S. Cristallo et al., Astroph.J. **859**, 105 (2018)
[4] C. Iliadis, *Nuclear Physics of Stars* (Weinheim, Germany, 2007)
[5] M. Busso et al., Astroph.J. **908**, 55 (2021)
[6] M. La Cognata et al., Astroph.J. **777**, 143 (2013)
[7] C. N. Davids, Nucl.Phys.A **110**, 619 (1968)
[8] J. K. Bair and F.X.Haas, Phys.Rev.C **7**, 1356 (1973)
[9] H. W. Drotleff et al., Astroph.J. **414**, 735 (1993)
[10] M. Heil et al.,Phys.Rev.C **78**, 025803 (2008)
[11] G. F. Ciani et al., Phys.Rev.Lett **127**, 152701 (2021)
[12] O. Trippella and M. L.Cognata, Astroph.J. **837**,41 (2017)
[13] A. J. Koning and M. C. Duijvestijn, Nucl. Phys. A **744**, 15 (2004)
[14] Y. Xu et al., Nucl. Phys. A. **918**, 61 (2013)
[15] W. Hauser and H. Feshbach, Phys.Rev. **87**, 366 (1952)
[16] S. Goriely et al., arXiv:0806.2239v1 [astro-ph] (2018)
[17] A. Gilbert and A. G. W. Cameron, Can. J. Phys., **43**, 1446 (1965)
[18] A. J. Koning and J. P. Delaroche, Nucl. Phys. A **713**, 231 (2003)
[19] V. Avrigeanu et al.,Phys. Rev.C **90**, 044612 (2014)
[20] E. D. Johnson et al., Phys. Rev. Lett **97**, 192701 (2006)
[21] C. Spitaleri et al., Eur. Phys. J. A. **55**, 161 (2019)
[22] R. G. Pizzone et al., Eur. Phys. J. A. **56**, 199 (2020)

Study for the long-lived gamma background due to neutron emitting calibration reactions

Tanmoy Bar[1,2,*], *Dipali* Basak[1,2], *Sukhendu* Saha[1,2], *Lalit Kumar* Sahoo[1,2], and *Chinmay* Basu[1]

[1]Saha Institute of Nuclear Physics, 1/AF Bidhannagar, Saltlake, Kolkata - 700064, India
[2]Homi Bhabha National Institute, 2nd floor, BARC Training School Complex, Anushaktinagar, Mumbai - 400094, India

Abstract. In this article, a detailed theoretical investigation has been done on the long-lived gamma background due to neutron emitting reactions. It mainly focused on the experiments that will be used to calibrate energy with the terminal voltage of an upcoming Facility for Research in Experimental Nuclear Astrophysics (FRENA). Many reactions like (p, n), (p, γ) have been utilized in various accelerator facilities around the globe for energy calibration purposes. Neutron emitting reactions like ^{7}Li(p, n), ^{13}C(p, n), ^{19}F(p, n), ^{27}Al(p, n), etc. have been very commonly used. For such reactions, a significant number of neutrons produced from such experiments can interact with surrounding elements like copper, tantalum, stainless steel (SS304 and SS316), concrete materials, etc. These interactions may create long-lived gamma activity in the vicinity of the accelerator and detection area. Background gammas from these radioactive isotopes can interfere with gamma measurements in future experiments. The present study has been done to investigate those possibilities of gamma background which may occur due to the calibration study by neutron emitting experiments. Here ^{7}Li(p, n), ^{13}C(p, n), ^{19}F(p, n) and ^{27}Al(p, n) reactions having neutron threshold energies 1.8803, 3.2355, 4.2351, 5.8036 MeV respectively are chosen keeping in mind the proton energy available at the facility.

1 Introduction

Calibrated energy is one of the most essential requirements of any accelerator. For this purpose, many experiments like threshold, resonance, etc. have been performed in various accelerator facilities. Neutron threshold experiments are one of such which have been widely used around the globe. The current theoretical study is mainly focused on the upcoming accelerator facility called FRENA [1]. It is a high current low energy (0.2 - 3) MV tandetron accelerator primarily dedicated to experiments related to nuclear astrophysics with very low cross-sections (~pb to nb). It is currently located at Saha Institute of Nuclear Physics, Kolkata, INDIA. This machine can deliver a proton beam with an energy range between 400 keV to 6 MeV with maximum beam intensity >300 μA. A systematic study is very important to know the probable gamma background during and due to experiments specially when a neutron is one of the major emitting particles. Neutrons can easily interact with the surrounding material and form isotopes which is radioactive. Isotopes forming from such interactions

*e-mail: tanmoy.bar@saha.ac.in

Table 1. Emitting neutron energy range at threshold energies for different reactions.

	^{7}Li(p, n)	^{13}C(p, n)	^{19}F(p, n)	^{27}Al(p, n)
Neutron energy range (keV)	20.09 - 41.12	11.20 - 23.41	4.39 - 19.91	3.05 - 14.00

having significantly large half-lives can create additional gamma background for experiments that will be performed exactly after that. This study is important for the radiation safety of accelerator workers as well.

2 Study of different neutron emitting reactions

In this study, ^{7}Li(p, n), ^{13}C(p, n), ^{19}F(p, n) and ^{27}Al(p, n) reactions having neutron threshold energies 1.8803, 3.2355, 4.2351, 5.8036 MeV respectively are used keeping in mind the proton energy availability at FRENA. Most of the components of the accelerators are made of 304 and 316 Stainless steel alloy along with copper (Cu) and Tantalum (Ta). Apart from that Aluminium (Al), Calcium (Ca), Silicon (Si), and Oxygen (O) is also present in the vicinity. All stable isotopes of these elements have been considered during the investigation. Reaction cross-section has been calculated using Hauser-Feshbach statistical model code TALYS version 1.95 [2] and emitting neutron energy range at threshold has been estimated using kinematic calculation. Kinematic calculation for ^{7}Li(p, n) is shown in Figure 1 and list for all reactions have been mentioned in Table 1.

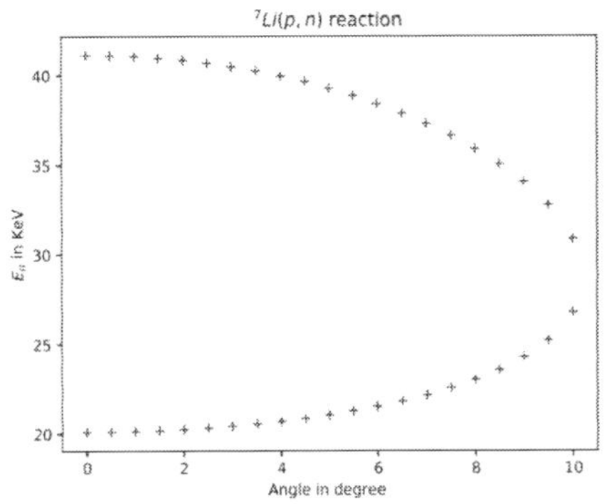

Figure 1. Neutron energy range of ^{7}Li(p, n) reaction at threshold energy (E_p = 1.8803 MeV).

3 Discussions on long lived activity

3.1 For proton at threshold energy

Neutrons emitting from these reactions at threshold energy as mentioned in Table 1 are within 50 keV. At this energy range, the most abundant tantalum isotope, ^{181}Ta (99.988%) is the only surrounding element that shows long-lived activity [3], having a half-life of more than 3 months. Fusion cross-section of ^{181}Ta with neutron in the energy range 10 - 40 keV is around ~ 10^{-3}mb. Gamma ray transition scheme is schematically shown in Figure 2.

^{181}Ta has half-live of ~114.74 days with most intense gamma rays of energies 1121.29 (35.24%), 1189.04 (16.49%), 1221.395 (27.23%), 1231.004 (11.62%) keV [4]. Complete list

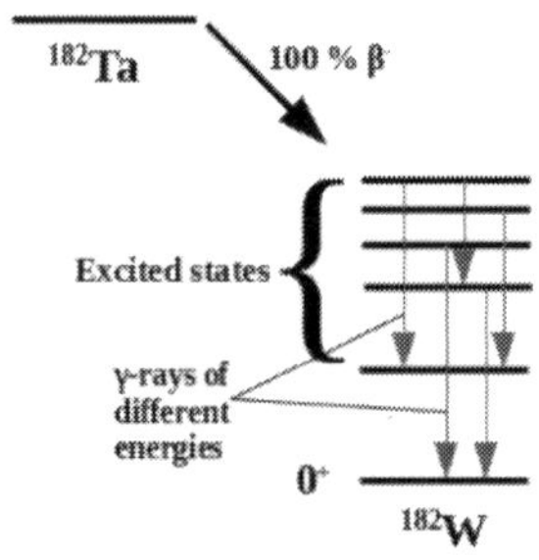

Figure 2. Schematic gamma transition scheme of ^{181}Ta.

of gamma having intensity greater than 1% along with cross-section value as calculated by TALYS for different neutron energy is shown in Table 2.

Table 2. List of gamma rays due to production of ^{182}Ta. Cross- section presented in the list is calculated by TALYS code and only gamma having intensity more than 1% are taken from NNDC [4].

Isotope	Abundance (%)	Neutron energy (keV)	Reaction product	Cross section (mb)	half life (days)	Daughter nucleus	γ-energy (keV)	γ intensity (%)
^{181}Ta	99.988	10	^{182}Ta	2.09×10^3	114.74	^{182}W	65.722	3.01
		20		1.34×10^3			67.749	42.9
		30		1.05×10^3			84.680	2.654
		40		8.69×10^2			100.106	14.2
							113.672	1.871
							152.429	7.02
							156.386	2.671
							179.393	3.119
							198.352	1.465
							222.109	7.57
							229.321	3.644
							264.074	3.612
							1001.7	2.086
							1121.29	35.24
							1189.04	16.49
							1221.395	27.23
							1231.004	11.62
							1257.407	1.51
							1289.145	1.37

3.2 For proton with higher energies

As the energy of the proton increases from the threshold value emitted neutron energy also gets increased, like for a 2 MeV proton, emitting neutrons will have energies between 16 - 232 keV. These energetic neutrons can interact with surrounding elements through different channels. In such a scenario, different isotopes will be produced which can contribute to the gamma background. As neutron energy goes beyond 350 keV, ^{54}Fe will produce ^{54}Mn having a half-life of 312 days. Most intense gamma-ray of ^{54}Mn is 834 keV (99.9%). Similarly, different isotopes will start to contribute as more and more reaction channels become accessible due to an increase in available neutron energy. Table 3 shows the contribution from different isotopes to the gamma background as neutron energy exceeds a specific value.

Table 3. Contribution of different isotopes due to increase in neutron energy. Isotopes in the vicinity and their natural abundance, produced isotopes after neutron capture and their half lives are listed in Columns 2 and 3. Column 4 and 5 indicated the final product and their most intense gamma rays along with relative intensity. Gamma energies and relative intensity are taken from NNDC [4].

Neutron Energy (keV)	Origin Isotope with Isotropic abundance (%)	Product after neutron capture with half life	Final product isotope	Most intense gamma rays(keV) with relative intensity (%)
> 350	^{54}Fe (5.85%)	^{54}Mn (312 days)	^{54}Cr	834 (99.9%)
> 400	^{58}Ni (68.07%)	^{58}Co (71 days)	^{58}Fe	810 (99.5%)
> 600	^{92}Mo (14.65%)	^{92}Nb (3.5 $\times 10^7$ years)	^{92}Zr	561 (100%),934 (74%)
> 850	^{40}Ca (96.94%)	^{40}K (1.3 $\times 10^9$ years)	^{40}Ar	1460 (10.66%)
> 900	^{50}Cr (4.35%)	^{50}V (2.6 $\times 10^{17}$ years)	^{50}Ti	1553 (99.3%)
> 1000	^{63}Cu (69.17%)	^{60}Co (5.27 years)	^{60}Ni	1173 (99.85%), 1332 (99.98%)
> 1600	^{98}Mo (24.29%)	^{95}Zr (64 days)	^{95}Nb	724 (44%), 756 (54%)
> 2500	^{94}Mo (9.19%)	^{94}Nb (2.1 $\times 10^4$ years)	^{94}Mo	702 (99.81%), 871 (99.89%)
> 3000	^{60}Ni (26.3%)	^{60}Co (5.27 years)	^{60}Ni	1173 (99.85%), 1332 (99.98%)

4 Conclusion

In this work, it was found that neutron emission occurs in the intermediate energy range for protons with energy close to the threshold value. For such reactions, only concerning element present in the nearby place is ^{181}Ta. But, in FRENA or any other accelerator facility tantalum is used mainly at the beam dump or in the faraday cup. Now, these components can be kept far away from the scattering region and can be shielded with lead bricks, High-Density Polyethylene (HDPE), and other conventional elements. In this way, detectors can be kept at a safe distance from gamma rays coming from the tantalum dump. Apart from tantalum, ^{58}Fe present in the visinity also form ^{59}Fe having half-life ~44 days with most intense gamma rays of energies 1099 keV(56.5%), 1291 keV(43.2%) etc. But the natural abundance of ^{58}Fe is only ~0.28%, so its contribution to the background may not be of much trouble. For reactions with protons at higher energy than the threshold, more isotopes need to be monitored as mentioned in Table 3.

References

[1] FRENA, *http : //www.saha.ac.in/web/frena − about − frena.*
[2] TALYS Code, *https : //tendl.web.psi.ch/tendl2019/talys.html*
[3] Tanmoy Bar et al., arXiv:2203.01995v1 [nucl-ex] 3 Mar 2022.
[4] National Nuclear Data Centre, https://www.nndc.bnl.gov/

First (p,n) reaction measurement in inverse kinematics with SECAR

Pelagia Tsintari[1,9,*], *Georg* Berg[3,10], *Jeff* Blackmon[6], *Kelly* Chipps[5], *Manoel* Couder[3,10], *Catherine* Deibel[6], *Nikolaos* Dimitrakopoulos[1,9], *Ruchi* Garg[2,8,9], *Uwe* Greife[7], *Kirby* Hermansen[2,8,9], *Ashley* Hood[6], *Rahul* Jain[2,8,9], *Cavan* Maher[2,8,9], *Caleb* Marshall[4,9], *Zach* Meisel[4,9], *Sara* Miskovich[2,8,9], *Fernando* Montes[2,9], *Georgios* Perdikakis[1,2,9], *Jorge* Pereira[2,9], *Thomas* Ruland[6], *Hendrik* Schatz[2,8,9], *Kiana* Setoodehnia[2], *Michael* Smith[5], *Louis* Wagner[2,9], and *Remco G. T.* Zegers[2,8,9]

[1]Department of Physics, Central Michigan University, Mt. Pleasant, MI 48859, USA
[2]Facility for Rare Isotope Beams, Michigan State University, East Lansing, MI 48824, USA
[3]Department of Physics and Astronomy, University of Notre Dame, Notre Dame, IN 46556, USA
[4]Department of Physics & Astronomy, Ohio University, Athens, OH 45701, USA
[5]Physics Division, Oak Ridge National Laboratory, Oak Ridge, TN 37831, USA
[6]Department of Physics and Astronomy, Louisiana State University, Baton Rouge, LA 70803, USA
[7]Department of Physics, Colorado School of Mines, Golden, CO 80401, USA
[8]Department of Physics and Astronomy, Michigan State University, East Lansing, MI 48824, USA
[9]The Joint Institute for Nuclear Astrophysics - Center for the Evolution of the Elements, Michigan State University, East Lansing, MI 48824, USA
[10]The Joint Institute for Nuclear Astrophysics - Center for the Evolution of the Elements, University of Notre Dame, Notre Dame, IN 46556, USA

Abstract. Nucleosynthesis in the νp-process occurs in regions of slightly proton-rich nuclei in the neutrino-driven wind of core-collapse supernovae. The process proceeds via a sequence of (p,γ) and (n,p) reactions, and depending on the conditions, may produce elements between Ni and Sn. Recent studies show that a few key (n,p) reactions regulate the efficiency of the neutrino-p process (νp-process). We performed a study of one of such (n,p) reactions via the measurement of the reverse (p,n) in inverse kinematics with SECAR at NSCL/FRIB. Such proton-induced reaction measurements are particularly challenging, as the recoils and the unreacted projectiles have nearly identical masses. An appropriate separation level can be achieved with SECAR, and along with the in-coincidence detection of neutrons these measurements become attainable. The preparation of the SECAR system for accommodating its first (p,n) reaction measurement, including the development of alternative ion beam optics, and the setup of the in-coincidence neutron detection, along with discussion on preliminary results from the p(^{58}Fe,n)^{58}Co reaction measurement are presented and discussed.

1 Introduction

Production of most of the stable neutron deficient isotopes has traditionally been attributed to the p-process [1]. Advances in the the modeling of the core collapse of Supernovae have

*e-mail: tsint1p@cmich.edu

shown that neutrino-driven winds immediately after the collapse may be proton-rich [2]. As such, a neutrino (νp) process [3] has been postulated as a possible source of abundances of those isotopes. The strong neutrino fluxes allow for electron anti-neutrino absorption on protons to produce a small residual neutron density. These neutrons are then captured via (n,p) reactions bypassing the typical reaction-flow slow-down due to the decay of nuclei with relatively long β-decay half-lives (such as the ^{56}Ni and ^{64}Ge) [4, 5]. Research efforts to develop a suitable technique to experimental constrain (n,p) and (p,n) reaction rates of short-lived nuclei [6] are underway at FRIB (Facility for Rare Isotope Beams). The measurement of the ^{58}Fe(p,n)^{58}Co reaction in inverse kinematics at SECAR (SEparator for CApture Reactions) described in this work is a recent effort in that direction.

2 The SEparator for CApture Reactions - SECAR

SECAR at NSCL/FRIB, consists of 8 dipole (B1-B8), 15 quadrupole (Q1-Q15), 3 hexapole (Hex1-Hex3), 1 (Oct1) octupole magnets, and 2 Wien filters (WF1-WF2). It has been designed with the required sensitivity to study (p,γ), and (a,γ) reactions, directly at astrophysical energies in inverse kinematics [7]. A first direct measurement of the known ^{58}Fe(p,n)^{58}Co reaction [8] was recently attained in inverse kinematics with SECAR and demonstrates that (p,n) cross section measurements are also possible with SECAR.

The ^{58}Fe beam delivered by ReA3/NSCL [9] at 3.8 MeV/u energy, with a charge state of 21^{+}, and an average intensity of 3.0E6 pps impinged on a polyethylene foil located at the SECAR target location [10, 11]. The delivered beam contained ^{58}Ni as a contaminant, estimated at around 20%. A C foil target was also used for background characterization. After a reaction in the target foil, the heavy ion recoils enter the SECAR system, along with the unreacted beam particles. The selection of a single charge state occurs using magnetic analysis in the first couple of dipoles (B1-B2). Two sets of a combination of a dipole pair and a velocity filter follow (B3-B4 and WF1, B5-B6 and WF2, respectively), providing the separation of the recoils from the unreacted projectiles for capture reactions. The final momentum analysis is performed with the last pair of dipole magnets (B7-B8) enhancing the rejection of scattered beam particles ("leaky beam"). The nearly identical masses between recoils and unreacted beam particles in the (p,n) reaction, pose a particular challenge and limit their separation solely to their energy difference. The energy difference between ^{58}Fe and ^{58}Co (as well as with the ^{58}Ni contaminant), mainly due to the reaction threshold, kinematics, and the stopping power effects (e.g. from target), is the only available dissimilarity to exploit via magnetic separation with SECAR.

3 Key components of the (p,n) technique development

Extensive ion-optical calculations were performed, using the COSY Infinity code [12] in multi-objective optimization routines to develop suitable ion optics for (p,n) measurements. The left and right panel at Figure 1 show the characteristic rays in the horizontal and the vertical plane, respectively. The heavy ions exiting the target foil have reached charge state equilibrium therefore, the most abundant charge state of ^{58}Co^{22+} recoils are shown in red, while the unreacted ^{58}Fe^{22+} particles are in blue (both set at the expected energy after the target). The key difference to the standard SECAR beam optics for the study of capture reactions (see Figure 3 in [7]) is the shift of the focal planes further downstream. This method enabled us to achieve sufficient magnetic separation at the position of the mass separation slits. Moreover, the Wien filters were inactive during the (p,n) measurement due to the fact that the difference in the magnetic rigidity of the beam and recoil ions was deemed adequate for their separation.

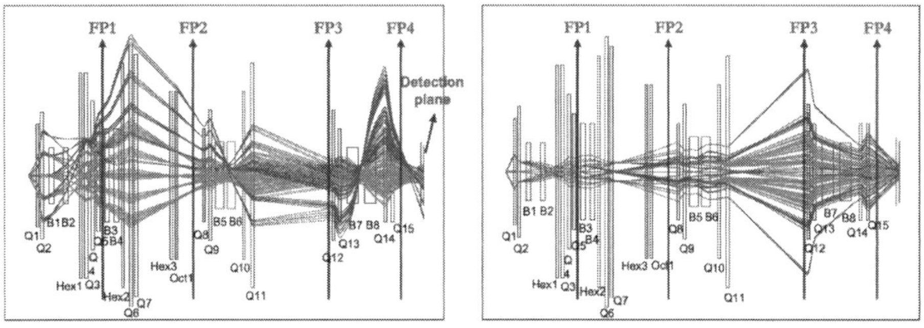

Figure 1. The ion optics of SECAR. The characteristic rays in the horizontal and vertical planes are shown in the left and right panel, respectively. The red lines represent the $^{58}Co^{22+}$ recoils, while the blue ones the unreacted $^{58}Fe^{22+}$ ions. The mass slits at every focal plane were used to cut down the majority of the unreacted beam.

As it can be observed in Figure 1, a small part of the unreacted beam particle distribution (blue lines) is not eliminated from reaching the final focal plane by the magnetic separation and the use of mass slits, yet it could still be discriminated with the use of in-coincidence detection systems. A gas ionization chamber (IC) and a double-sided silicon strip detector (DSSD), installed at the detection plane at the end of SECAR beam-line, provided a ΔE-E particle identification. A new installation of a neutron detection system, at the start of the SECAR beam-line, close to the target location, included four organic liquid scintillators (EJ-301, cylindrical 2x2 inches active volume) in a ring configuration, and 21 LENDA bars (Low-Energy Neutron Detection Array) [13]. These neutron detectors provide crucial neutron-tagging and permit the event-by-event in-coincidence detection between neutrons emitted at the target location and recoils reaching the SECAR final focal plane. The two planar silicon surface barrier detectors of 75 μm thickness and 300 mm^2 active area, located inside the target chamber were used for monitoring beam intensity, via the detection of scattered target particles.

4 Discussion

A preliminary analysis demonstrates that the percentage of the unreacted beam particles and "leaky beam" reaching the detection plane of SECAR does not exceed 0.01%. Therefore, this performance is expected to be sufficient for separating the beam and recoil ions using in-coincidence the neutron and focal plane detection systems. As the analysis is still ongoing, correction factors such as neutron efficiency, transmission, rejection and acceptance of the system, are yet to be determined. The completed work is expected to be presented in a future publication.

References

[1] G. Wallerstein, et al., Rev. Mod. Phys., **69**, 995-1084, (1997)
[2] Fischer, et al., Astronomy and Astrophysics, **517**, A80+, (2010)
[3] C. Froehlich, et al., Phys. Rev. Lett., **96**, 142507, (2006)
[4] S. Wanajo, et al., APJ, **729**, 46, (2011)
[5] A. Arcones, et al., APJ, **750**, 18, (2012)

[6] P. Gastis, et al., Nucl. Inst. Meth. Phys. Res. A, **985**, 164603, (2021)
[7] G. P. A. Berg, et al., Nucl. Inst. Meth. Phys. Res. A, **877**, 87–103 (2018)
[8] S.G. Tims, et al., Nucl. Inst. Meth. Phys. Res. A, **563**, 473-493, (1993)
[9] D. Leitner, Nucl. Inst. Meth. Phys. Res. B., **317**, 235-241, (2013)
[10] K.A. Chipps, et al., Nucl. Inst. Meth. Phys. Res. A, **763**, 553–564 (2014)
[11] K. Schmidt, et al., Nucl. Inst. Meth. Phys. Res. A, **911**, 1-9 (2018)
[12] M. Berz, COSY Infinity, $http://www.bt.pa.msu.edu/index_files/cosy.html$
[13] G. Perdikakis, et al., Nucl. Inst. Meth. Phys. Res. A, **686**, 117-124, (2012)

AUTHOR INDEX

AUTHOR INDEX

AUTHOR INDEX

EDP Sciences – Web of Conferences
17, Avenue du Hoggar
Parc d'Activité de Courtabouf
BP 112
F-91944 Les Ulis Cedex A
France

ISBN 978-1-7138-7049-4

CURRAN ASSOCIATES INC.
proceedings.com

9 781713 870494